Energy Conservation

Energy Conservation

Edited by
Daniel Walker

Larsen & Keller
www.larsen-keller.com

Energy Conservation
Edited by Daniel Walker
ISBN: 978-1-63549-103-6 (Hardback)

Larsen & Keller

Published by Larsen and Keller Education,
5 Penn Plaza,
19th Floor,
New York, NY 10001, USA

Cataloging-in-Publication Data

Energy conservation / edited by Daniel Walker.
p. cm.
Includes bibliographical references and index.
ISBN 978-1-63549-103-6
1. Energy conservation. 2. Power resources. 3. Renewable energy sources.
4. Energy industries--Environmental aspects. I. Walker, Daniel.

TJ163.3 . E54 2017
621.042--dc23

Printed and bound in China.

For more information regarding Larsen and Keller Education and its products, please visit the publisher's website www.larsen-keller.com

Table of Contents

Preface

Due to the rise in world population in the last couple of decades, the need for energy conservation has dramatically risen. Governments across the globe are highlighting the use of alternative energy production methods like solar panels, hydropower, wind energy, etc. so as to limit the damage done to the environment. The aim of this text is to equip students with all the latest developments made in this field. The topics introduced in the book cover all the elements from the most basic to the most complex. Those in search of information to further their knowledge will be greatly assisted by this book. As this field is emerging at a rapid pace, the contents of this textbook will help the readers understand the modern concepts and applications of the subject.

A detailed account of the significant topics covered in this book is provided below:

Chapter 1- The conservation of energy by using less energy is known as energy conservation. It differs from efficient energy use. Some examples of energy conservation are unplugging your computers or electronic devices when they are not in use or avoiding driving vehicles. In both these examples, what is common is that we are reducing the use of energy.

Chapter 2- Sustainable energy is the energy that is used in the present without negotiating the needs of the future. Renewable energy, renewable energy commercialization and energy storage are some of the themes of this text. This chapter has been carefully written to provide an easy understanding of the varied facets of sustainable energy.

Chapter 3- Efficient energy use is the aim of reducing the usage of energy and provides products that help the same, for example, the usage of natural skylights. The section focuses on other aspects of efficient use such as energy efficiency in transportation, energy efficiency implementation, energy intensity and energy efficiency gap.

Chapter 4- New technologies have increased the efficiency of resource use and the benefits caused by these technologies are known as the rebound effect. Rebound effect, megawatt power, Earth hour, energy hierarchy and one-watt initiative are some of the key movements of energy conservation. The following chapter unfolds its crucial aspects in a critical yet systematic manner.

Chapter 5- Energy efficient landscaping is designed to preserve energy. Some of the examples of energy efficient landscaping are planting of trees, earth sheltering and planting of windbreaks. This segment expounds topics such as low energy house, green building, passive house and passive cooling.

Chapter 6- The applications of energy conservation are renewable heat, smart grid, thermal energy storage and home energy saver. Renewable heat is a utilization of renewable energy. The heat technologies involved are solar heating, geothermal heating, heat pumps and heat exchangers. The aspects elucidated in the chapter are of vital importance, and provide a better understanding of energy conservation.

Chapter 7- Humans have used energy for hundreds of years. We have consumed natural resources to the extent that it has caused global warming. Changes in most parts of the worlds can already be noticed. The following section will not only provide an overview but it also delves into the topics related to it.

I would like to make a special mention of my publisher who considered me worthy of this opportunity and also supported me throughout the process. I would also like to thank the editing team at the back-end who extended their help whenever required.

Editor

1

An Introduction to Energy Conservation

The conservation of energy by using less energy is known as energy conservation. It differs from efficient energy use. Some examples of energy conservation are unplugging your computers or electronic devices when they are not in use or avoiding driving vehicles. In both these examples, what is common is that we are reducing the use of energy.

Energy Conservation

Energy conservation refers to reducing energy consumption through using less of an energy service. Energy conservation differs from efficient energy use, which refers to using less energy for a constant service. Driving less is an example of energy conservation. Driving the same amount with a higher mileage vehicle is an example of energy efficiency. Energy conservation and efficiency are both energy reduction techniques. Energy conservation is a part of the concept of sufficiency.

Even though energy conservation reduces energy services, it can result in increased environmental quality, national security, personal financial security and higher savings. It is at the top of the sustainable energy hierarchy. It also lowers energy costs by preventing future resource depletion.

Energy Tax

Some countries employ energy or carbon taxes to motivate energy users to reduce their consumption. Carbon taxes can allow consumption to shift to nuclear power and other alternatives that carry a different set of environmental side effects and limitations. Meanwhile, taxes on all energy consumption stand to reduce energy use across the board, while reducing a broader array of environmental consequences arising from energy production. The State of California employs a tiered energy tax whereby every consumer receives a baseline energy allowance that carries a low tax. As usage increases above that baseline, the tax is increasing drastically. Such programs aim to protect poorer households while creating a larger tax burden for high energy consumers.

Building Design

One of the primary ways to improve energy conservation in buildings is to use an energy audit. An energy audit is an inspection and analysis of energy use and flows for energy conservation in a building, process or system to reduce the amount of energy input into the system without negatively affecting the output(s). This is normally accomplished by trained professionals and can be part of some of the national programs discussed above. In addition, recent development of smartphone apps enable homeowners to complete relativily sophisticated energy audits themselves.

Building technologies and smart meters can allow energy users, business and residential, to see graphically the impact their energy use can have in their workplace or homes. Advanced real-time energy metering is able to help people save energy by their actions.

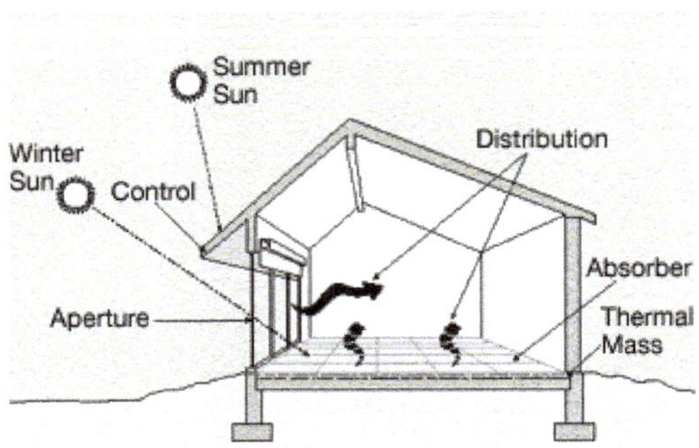

Elements of passive solar design, shown in a direct gain application

In passive solar building design, windows, walls, and floors are made to collect, store, and distribute solar energy in the form of heat in the winter and reject solar heat in the summer. This is called passive solar design or climatic design because, unlike active solar heating systems, it doesn't involve the use of mechanical and electrical devices.

The key to designing a passive solar building is to best take advantage of the local climate. Elements to be considered include window placement and glazing type, thermal insulation, thermal mass, and shading. Passive solar design techniques can be applied most easily to new buildings, but existing buildings can be retrofitted.

Transportation

In the United States, suburban infrastructure evolved during an age of relatively easy access to fossil fuels, which has led to transportation-dependent systems of living. Zoning reforms that allow greater urban density as well as designs for walking and bicycling can greatly reduce energy consumed for transportation. The use of telecommuting by major corporations is a significant opportunity to conserve energy, as many Americans now work in service jobs that enable them to work from home instead of commuting to work each day.

Consumer Products

Consumers are often poorly informed of the savings of energy efficient products. A prominent example of this is the energy savings that can be made by replacing an incandescent light bulb with a more modern alternative. When purchasing light bulbs, many consumers opt for cheap incandescent bulbs, failing to take into account their higher energy costs and lower lifespans when compared to modern compact fluorescent and LED bulbs. Although these energy-efficient alternatives have a higher upfront cost, their long lifespan and low energy use can save consumers a considerable amount of money. The price of LEDs has also been steadily decreasing in the past five years, due to improvement of the semiconductor technology. Many LED bulbs on the market qualify for utility rebates that further reduce the price of purchase to the consumer. Estimates by the U.S. Department of Energy state that widespread adoption of LED lighting over the next 20 years could result in about $265 billion worth of savings in United States energy costs.

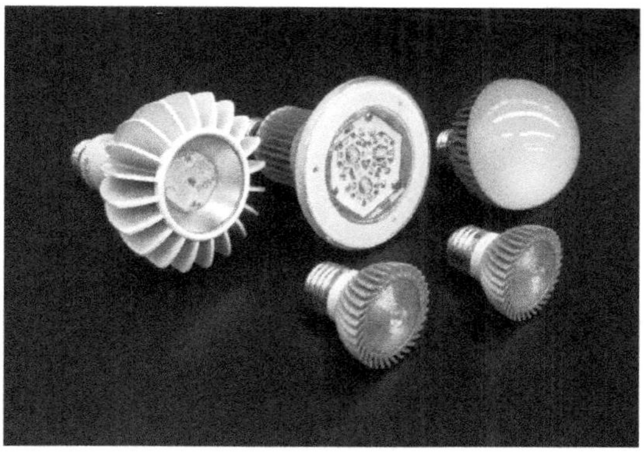

An assortment of energy-efficient semiconductor (LED) lamps for commercial and residential lighting use. LED lamps use at least 75% less energy, and last 25 times longer, than traditional incandescent light bulbs.

The research one must put into conserving energy is often too time consuming and costly for the average consumer, when there are cheaper products and technology available using today's fossil fuels. Some governments and NGOs are attempting to reduce this complexity with ecolabels that make differences in energy efficiency easy to research while shopping.

To provide the kind of information and support people need to invest money, time and effort in energy conservation, it is important to understand and link to people's topical concerns. For instance, some retailers argue that bright lighting stimulates purchasing. However, health studies have demonstrated that headache, stress, blood pressure, fatigue and worker error all generally increase with the common over-illumination present in many workplace and retail settings. It has been shown that natural daylighting increases productivity levels of workers, while reducing energy consumption.

In warm climates where air conditioning is used, any household device that gives off heat will result in a larger load on the cooling system. Items such as a stove, dish washer, clothes dryer, hot water and incandescent lighting all add heat to the home. Low power or insulated versions of these devices give off less heat for the air conditioning to remove. The air conditioning system can also improve in efficiency by using a heat sink that is cooler than the standard air heat exchanger such as geothermal or water.

In cold climates heating air and water is a major demand on household energy use. By investing in newer technologies in the home, significant energy reductions are possible. Heat pumps are a more efficient alternative to using electrical resistance heaters for warming air or water. A variety of efficient clothes dryers are available, and the classic clothes line requires no energy, only time. Natural gas condensing boilers and hot air furnaces increase efficiency over standard hot flue models. New construction implementing heat exchangers can capture heat from waste water or exhaust air in bathrooms, laundry and kitchens.

In both warm and cold climate extremes, airtight thermal insulated construction will largely determine the efficiency of a home. Insulation is added to minimize the flow of heat to or from the home, but can be labor-intensive to retrofit to an existing home.

Energy Conservation By the Countries

Asia

Despite the vital role energy efficiency is envisaged to play in cost-effectively cutting energy demand, only a small part of its economic potential is exploited in the Asia. Governments have implemented a range of subsidies such as cash grants, cheap credit, tax exemptions, and co-financing with public-sector funds to encourage a range of energy-efficiency initiatives across several sectors. Governments in the Asia-Pacific region have implemented a range of information provision and labeling programs for buildings, appliances, and the transportation and industrial sectors. Information programs can simply provide data, such as fuel-economy labels, or actively seek to encourage behavioral changes, such as Japan's Cool Biz campaign that encourages setting air conditioners at 28-degrees Celsius and allowing employees to dress casually in the summer. More in Pacific Energy Summit.

European Union

At the end of 2006, the European Union (EU) pledged to cut its annual consumption of primary energy by 20% by 2020. The 'European Union Energy Efficiency Action Plan' is long awaited. Directive 2012/27/EU is on energy efficiency.

As part of the EU's SAVE Programme, aimed at promoting energy efficiency and encouraging energy-saving behaviour, the Boiler Efficiency Directive specifies minimum levels of efficiency for boilers fired with liquid or gaseous fuels.

India

Petroleum Conservation Research Association (PCRA) is an Indian government body created in 1977 and engaged in promoting energy efficiency and conservation in every walk of life. In the recent past PCRA has done mass media campaigns in television, radio & print media. An impact assessment survey by a third party revealed that due to these mega campaigns by PCRA, overall awareness level have gone up leading to saving of fossil fuels worth crores of rupees(Indian currency) besides reducing pollution.

Bureau of Energy Efficiency is an Indian governmental organization created in 2001 responsible for promoting energy efficiency and conservation.

Japan

Advertising with high energy in Shinjuku, Japan.

Since the 1973 oil crisis, energy conservation has been an issue in Japan. All oil based fuel is imported, so indigenous sustainable energy is being developed.

The Energy Conservation Center promotes energy efficiency in every aspect of Japan. Public entities are implementing the efficient use of energy for industries and research. It includes projects such as the Top Runner Program. In this project, new appliances are regularly tested on efficiency, and the most efficient ones are made the standard.

Lebanon

In Lebanon and since 2002 The Lebanese Center for Energy Conservation (LCEC) has been promoting the development of efficient and rational uses of energy and the use of renewable energy at the consumer level. It was created as a project financed by the International Environment Facility (GEF) and the Ministry of Energy Water (MEW) under the management of the United Nations Development Programme (UNDP) and gradually established itself as an independent technical national center although it continues to be supported by the United Nations Development Programme (UNDP) as indicated in the Memorandum of Understanding (MoU) signed between MEW and UNDP on June 18, 2007.

Nepal

Until recently, Nepal has been focusing on the exploitation of its huge water resources to produce hydro power. Demand side management and energy conservation was not in the focus of government action. In 2009, bilateral Development Cooperation between Nepal and the Federal Republic of Germany, has agreed upon the joint implementation of “Nepal Energy Efficiency Programme”. The lead executing agencies for the implementation are the Water and Energy Commission Secretariat (WECS). The aim of the programme is the promotion of energy efficiency in policy making, in rural and urban households as well as in the industry. Due to the lack of a government organization that promotes energy efficiency in the country, the Federation of Nepalese Chambers of Commerce and Industry (FNCCI) has established the Energy Efficiency Centre under his roof to promote energy conservation in the private sector. The Energy Efficiency Centre is a non-profit initiative that is offering energy auditing services to the industries. The Centre is also supported by Nepal Energy Efficiency Programme of Deutsche Gesellschaft für Internationale Zusammenarbeit. A study conducted in 2012 found out that Nepalese industries could save 160,000 Megawatt hours of electricity and 8,000 Terajoule of thermal energy (like diesel, furnace oil and coal) every year. These savings are equivalent to annual energy cost cut of up to 6.4 Billion Nepalese Rupees. As a result of Nepal Economic Forum 2014, an economic reform agenda in the priority sectors was declared focusing on energy conservation among others. In the energy reform agenda the government of Nepal gave the commitment to introduce incentive packages in the budget of the fiscal year 2015/16 for industries that practices energy efficiency or use efficient technologies (incl. cogeneration).

New Zealand

In New Zealand the Energy Efficiency and Conservation Authority is the Government Agency responsible for promoting energy efficiency and conservation. The Energy Management Association of New Zealand is a membership based organization representing the New Zealand energy services sector, providing training and accreditation services with the aim of ensuring energy management services are credible and dependable.

Nigeria

In Nigeria, the Lagos State Government is encouraging Lagosians to imbibe an energy conservation culture. The Lagos State Electricity Board (LSEB) is spearheading an initiative tagged “Conserve Energy, Save Money” under the Ministry of Energy and Mineral Resources. The initiative is designed to sensitize Lagosians around the theme of energy conservation by connecting with and influencing their behavior through do-it-yourself tips and exciting interaction with prominent personalities. In September 2013, Governor Babatunde Raji Fashola of Lagos State and rapper Jude ‘MI’ Abaga (campaign ambassador)() participated in the Governor’s first ever Google+ Hangout on YouTube on the topic of energy conservation.

In addition to the hangout, during the month of October (the official energy conservation month in the state), LSEB hosted experience centers in malls around Lagos State where members of the public were encouraged to calculate their current household energy consumption and discover ways to save money using the 1st-ever consumer-focused energy app in sub-saharan Africa (). To get Lagosians started on energy conservation, Solar Lamps and Phillips Energy-saving bulbs were also given out at each experience center. Pictures from the experience centers: (part of Lagos state government energy initiatives)

Sri Lanka

Sri Lanka currently consumes fossil fuels, hydro power, wind power, solar power and dendro power for their day to day power generation. The Sri Lanka Sustainable Energy Authority is playing a major role regarding energy management and energy conservation. Today, most of the industries are requested to reduce their energy consumption by using renewable energy sources and optimizing their energy usage.

Turkey

Turkey aims to decrease by at least 20% the amount of energy consumed per GDP of Turkey by the year 2023 (energy intensity).

United States

The United States is currently the second largest single consumer of energy, following China. The U.S. Department of Energy categorizes national energy use in four broad sectors: transportation, residential, commercial, and industrial.

Energy usage in transportation and residential sectors, about half of U.S. energy consumption, is largely controlled by individual consumers. Commercial and industrial energy expenditures are determined by businesses entities and other facility managers. National energy policy has a significant effect on energy usage across all four sectors.

References

- "Key concepts: energy efficiency: energy conservation vs. energy efficiency". Natural Resources Canada. Retrieved 2016-06-07.
- Zehner, Ozzie (2011). Unintended consequences of green technologies — Uncorrected proof (PDF). Retrieved 2016-06-14.
- "Baseline Study of Selected Sector Industries to assess The Potentials for more Efficient use of Energy in Nepal" (PDF). Deutsche Gesellschaft für Internationale Zusammenarbeit (GIZ). 2016. Retrieved 2 January 2014.
- "NEPAL ECONOMIC SUMMIT 2014 DECLARATION – A COMMITMENT TO ECONOMIC REFORM". FNCCI. 2014. Retrieved 7 April 2014.
- "Baseline Study of Selected Sector Industries to assess The Potentials for more Efficient use of Energy in Nepal". Deutsche Gesellschaft für Internationale Zusammenarbeit (GIZ). 2012. Retrieved 2 January 2014.
- "Nepal Energy Efficiency Programme". Water and Energy Commission Secretariat (WECS). 2011. Retrieved 6 December 2013.

2

Understanding Sustainable Energy

Sustainable energy is the energy that is used in the present without negotiating the needs of the future. Renewable energy, renewable energy commercialization and energy storage are some of the themes of this text. This chapter has been carefully written to provide an easy understanding of the varied facets of sustainable energy.

Sustainable Energy

Sustainable energy is energy that is consumed at insignificant rates compared to its supply and with manageable collateral effects, especially environmental effects. Another common definition of sustainable energy is an energy system that serves the needs of the present without compromising the ability of future generations to meet their needs. The organizing principle for sustainability is sustainable development, which includes the four interconnected domains: ecology, economics, politics and culture. Sustainability science is the study of sustainable development and environmental science.

Technologies that promote sustainable energy include renewable energy sources, such as hydroelectricity, solar energy, wind energy, wave power, geothermal energy, bioenergy, tidal power and also technologies designed to improve energy efficiency. Costs have fallen dramatically in recent years, and continue to fall. Most of these technologies are either economically competitive or close to being so. Increasingly, effective government policies support investor confidence and these markets are expanding. Considerable progress is being made in the energy transition from fossil fuels to ecologically sustainable systems, to the point where many studies support 100% renewable energy.

Definitions

Energy efficiency and renewable energy are said to be the *twin pillars* of sustainable energy. Some ways in which *sustainable energy* has been defined are:

- "Effectively, the provision of energy such that it meets the needs of the present without compromising the ability of future generations to meet their own needs. ...Sustainable Energy has two key components: renewable energy and energy efficiency." – *Renewable Energy and Efficiency Partnership* (British)
- "Dynamic harmony between equitable availability of energy-intensive goods and services to all people and the preservation of the earth for future gener-

ations." And, "The solution will lie in finding sustainable energy sources and more efficient means of converting and utilizing energy." – *Sustainable Energy* by J. W. Tester, *et al.*, from MIT Press.

- "Any energy generation, efficiency and conservation source where: Resources are available to enable massive scaling to become a significant portion of energy generation, long term, preferably 100 years.." – *Invest,* a green technology non-profit organization.
- "Energy which is replenishable within a human lifetime and causes no long-term damage to the environment." – *Jamaica Sustainable Development Network*

This sets *sustainable energy* apart from other renewable energy terminology such as *alternative energy* by focusing on the ability of an energy source to continue providing energy. Sustainable energy can produce some pollution of the environment, as long as it is not sufficient to prohibit heavy use of the source for an indefinite amount of time. Sustainable energy is also distinct from low-carbon energy, which is sustainable only in the sense that it does not add to the CO_2 in the atmosphere.

Green Energy is energy that can be extracted, generated, and/or consumed without any significant negative impact to the environment. The planet has a natural capability to recover which means pollution that does not go beyond that capability can still be termed green.

Green power is a subset of renewable energy and represents those renewable energy resources and technologies that provide the highest environmental benefit. The U.S. Environmental Protection Agency defines green power as electricity produced from solar, wind, geothermal, biogas, biomass and low-impact small hydroelectric sources. Customers often buy green power for avoided environmental impacts and its greenhouse gas reduction benefits.

Renewable Energy Technologies

Renewable energy technologies are essential contributors to sustainable energy as they generally contribute to world energy security, reducing dependence on fossil fuel resources, and providing opportunities for mitigating greenhouse gases. The International Energy Agency states that:

Conceptually, one can define three generations of renewables technologies, reaching back more than 100 years .

First-generation technologies emerged from the industrial revolution at the end of the 19th century and include hydropower, biomass combustion and geothermal power and heat. Some of these technologies are still in widespread use.

Second-generation technologies include solar heating and cooling, wind power, modern forms of bioenergy and solar photovoltaics. These are now entering markets as a result of research, development and demonstration (RD&D) investments since the 1980s. The initial investment was prompted by energy security concerns linked to the oil crises (1973 and 1979) of the 1970s but the continuing appeal of these renewables is due, at least in part, to environmental benefits. Many of the technologies reflect significant advancements in materials.

Third-generation technologies are still under development and include advanced biomass gasification, biorefinery technologies, concentrating solar thermal power, hot dry rock geothermal energy and ocean energy. Advances in nanotechnology may also play a major role.

—International Energy Agency, RENEWABLES IN GLOBAL ENERGY SUPPLY, An IEA Fact Sheet

First- and second-generation technologies have entered the markets, and third-generation technologies heavily depend on long term research and development commitments, where the public sector has a role to play.

Regarding energy used by vehicles, a comprehensive 2008 cost-benefit analysis review was conducted of sustainable energy sources and usage combinations in the context of global warming and other dominating issues; it ranked wind power generation combined with battery electric vehicles (BEV) and hydrogen fuel cell vehicles (HFCVs) as the most efficient. Wind was followed by concentrated solar power (CSP), geothermal power, tidal power, photovoltaic, wave power, hydropower coal capture and storage (CCS), nuclear energy and biofuel energy sources. It states: "In sum, use of wind, CSP, geothermal, tidal, PV, wave, and hydro to provide electricity for BEVs and HFCVs and, by extension, electricity for the residential, industrial, and commercial sectors, will result in the most benefit among the options considered. The combination of these technologies should be advanced as a solution to global warming, air pollution, and energy security. Coal-CCS and nuclear offer less benefit thus represent an opportunity cost loss, and the biofuel options provide no certain benefit and the greatest negative impacts."

First-generation Technologies

One of many power plants at The Geysers, a geothermal power field in northern California, with a total output of over 750 MW.

First-generation technologies are most competitive in locations with abundant resources. Their future use depends on the exploration of the available resource potential, particularly in developing countries, and on overcoming challenges related to the environment and social acceptance.

—International Energy Agency, RENEWABLES IN GLOBAL ENERGY SUPPLY, An IEA Fact Sheet

Among sources of renewable energy, hydroelectric plants have the advantages of being long-lived—many existing plants have operated for more than 100 years. Also, hydroelectric plants are clean and have few emissions. Criticisms directed at large-scale hydroelectric plants include: dislocation of people living where the reservoirs are planned, and release of significant amounts of carbon dioxide during construction and flooding of the reservoir.

Hydroelectric dams are one of the most widely deployed sources of sustainable energy.

However, it has been found that high emissions are associated only with shallow reservoirs in warm (tropical) locales, and recent innovations in hydropower turbine technology are enabling efficient development of low-impact run-of-the-river hydroelectricity projects. Generally speaking, hydroelectric plants produce much lower life-cycle emissions than other types of generation. Hydroelectric power, which underwent extensive development during growth of electrification in the 19th and 20th centuries, is experiencing resurgence of development in the 21st century. The areas of greatest hydroelectric growth are the booming economies of Asia. China is the development leader; however, other Asian nations are installing hydropower at a rapid pace. This growth is driven by much increased energy costs—especially for imported energy—and widespread desires for more domestically produced, clean, renewable, and economical generation.

Geothermal power plants can operate 24 hours per day, providing base-load capacity, and the world potential capacity for geothermal power generation is estimated at 85 GW over the next 30 years. However, geothermal power is accessible only in limited areas of the world, including the United States, Central America, East Africa, Iceland, Indonesia, and the Philippines. The costs of geothermal energy have dropped substantially from the systems built in the 1970s. Geothermal heat generation can be competitive in many countries producing geothermal power, or in other regions where the resource is of a lower temperature. Enhanced geothermal system (EGS) technology does not require natural convective hydrothermal resources, so it can be used in areas that were previously unsuitable for geothermal power, if the resource is very large. EGS is currently under research at the U.S. Department of Energy.

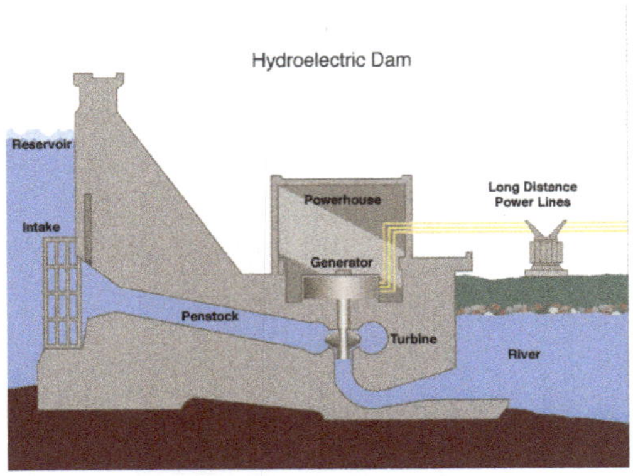

Hydroelectric dam in cross section

Biomass briquettes are increasingly being used in the developing world as an alternative to charcoal. The technique involves the conversion of almost any plant matter into compressed briquettes that typically have about 70% the calorific value of charcoal. There are relatively few examples of large-scale briquette production. One exception is in North Kivu, in eastern Democratic Republic of Congo, where forest clearance for charcoal production is considered to be the biggest threat to mountain gorilla habitat. The staff of Virunga National Park have successfully trained and equipped over 3500 people to produce biomass briquettes, thereby replacing charcoal produced illegally inside the national park, and creating significant employment for people living in extreme poverty in conflict-affected areas.

Second-generation Technologies

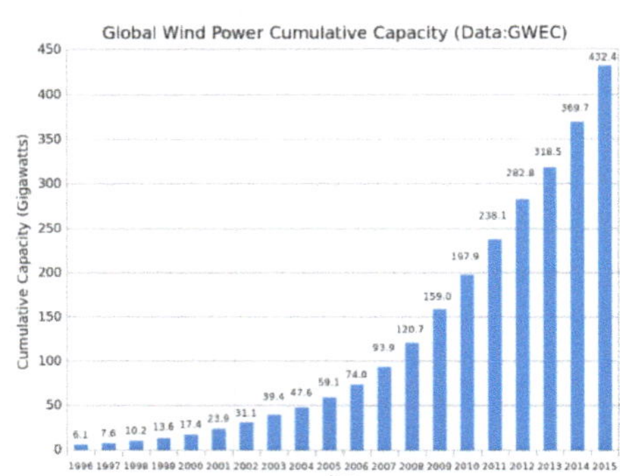

Wind power: worldwide installed capacity

Markets for second-generation technologies are strong and growing, but only in a few countries. The challenge is to broaden the market base for continued growth worldwide. Strategic deployment in one country not only reduces technology costs for users there, but also for those in other countries, contributing to overall cost reductions and performance improvement.

—International Energy Agency, RENEWABLES IN GLOBAL ENERGY SUPPLY, An IEA Fact Sheet

Solar heating systems are a well known second-generation technology and generally consist of solar thermal collectors, a fluid system to move the heat from the collector to its point of usage, and a reservoir or tank for heat storage and subsequent use. The systems may be used to heat domestic hot water, swimming pool water, or for space heating. The heat can also be used for industrial applications or as an energy input for other uses such as cooling equipment. In many climates, a solar heating system can provide a very high percentage (50 to 75%) of domestic hot water energy. Energy received from the sun by the earth is that of electromagnetic radiation. Light ranges of visible, infrared, ultraviolet, x-rays, and radio waves received by the earth through solar energy. The highest power of radiation comes from visible light. Solar power is complicated due to changes in seasons and from day to night. Cloud cover can also add to complications of solar energy, and not all radiation from the sun reaches earth because it is absorbed and dispersed due to clouds and gases within the earth's atmospheres.

11 MW solar power plant near Serpa, Portugal 38°1′51″N 7°37′22″W38.03083°N 7.62278°W

In the 1980s and early 1990s, most photovoltaic modules provided remote-area power supply, but from around 1995, industry efforts have focused increasingly on developing building integrated photovoltaics and power plants for grid connected applications.Currently the largest photovoltaic power plant in North America is the Nellis Solar Power Plant (15 MW). There is a proposal to build a Solar power station in Victoria, Australia, which would be the world's largest PV power station, at 154 MW. Other large photovoltaic power stations include the Girassol solar power plant (62 MW), and the Waldpolenz Solar Park (40 MW).

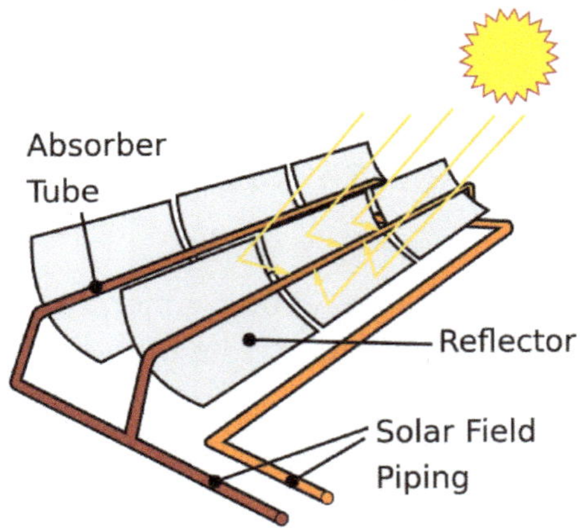

Sketch of a Parabolic Trough Collector

Some of the second-generation renewables, such as wind power, have high potential and have already realised relatively low production costs. At the end of 2008, worldwide wind farm capacity was 120,791 megawatts (MW), representing an increase of 28.8 percent during the year, and wind power produced some 1.3% of global electricity consumption. Wind power accounts for approximately 20% of electricity use in Denmark, 9% in Spain, and 7% in Germany. However, it may be difficult to site wind turbines in some areas for aesthetic or environmental reasons, and it may be difficult to integrate wind power into electricity grids in some cases.

Solar thermal power stations have been successfully operating in California commercially since the late 1980s, including the largest solar power plant of any kind, the 350 MW Solar Energy Generating Systems. Nevada Solar One is another 64MW plant which has recently opened. Other parabolic trough power plants being proposed are two 50MW plants in Spain, and a 100MW plant in Israel.

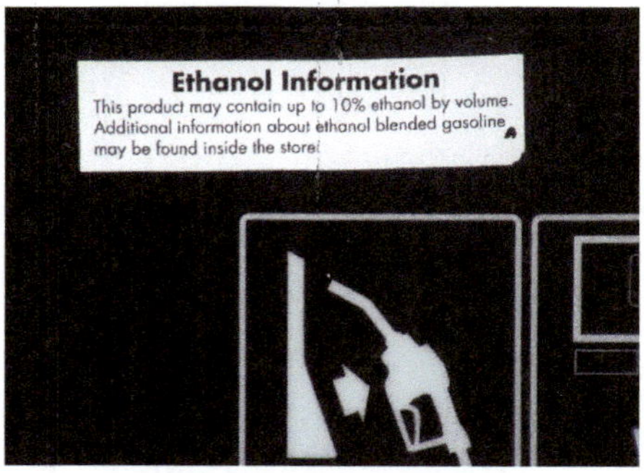

Information on pump, California

Brazil has one of the largest renewable energy programs in the world, involving production of ethanol fuel from sugar cane, and ethanol now provides 18 percent of the country's automotive fuel. As a result of this, together with the exploitation of domestic deep water oil sources, Brazil, which years ago had to import a large share of the petroleum needed for domestic consumption, recently reached complete self-sufficiency in oil.

Most cars on the road today in the U.S. can run on blends of up to 10% ethanol, and motor vehicle manufacturers already produce vehicles designed to run on much higher ethanol blends. Ford, DaimlerChrysler, and GM are among the automobile companies that sell "flexible-fuel" cars, trucks, and minivans that can use gasoline and ethanol blends ranging from pure gasoline up to 85% ethanol (E85). By mid-2006, there were approximately six million E85-compatible vehicles on U.S. roads.

Third-generation Technologies

MIT's Solar House#1 built in 1939 used seasonal thermal energy storage (STES) for year-round heating.

Third-generation technologies are not yet widely demonstrated or commercialised. They are on the horizon and may have potential comparable to other renewable energy technologies, but still depend on attracting sufficient attention and RD&D funding. These newest technologies include advanced biomass gasification, biorefinery technologies, solar thermal power stations, hot dry rock geothermal energy and ocean energy.

—*International Energy Agency, RENEWABLES IN GLOBAL ENERGY SUPPLY, An IEA Fact Sheet*

Bio-fuels may be defined as "renewable," yet may not be "sustainable," due to soil degradation. As of 2012, 40% of American corn production goes toward ethanol. Ethanol takes up a large percentage of "Clean Energy Use" when in fact, it is still debatable whether ethanol should be considered as a "Clean Energy."

According to the International Energy Agency, new bioenergy (biofuel) technologies being developed today, notably cellulosic ethanol biorefineries, could allow biofuels to play a much bigger role in the future than previously thought. Cellulosic ethanol can be made from plant matter composed primarily of inedible cellulose fibers that form the stems and branches of most plants. Crop residues (such as corn stalks, wheat straw and rice straw), wood waste and municipal solid waste are potential sources of cellulosic biomass. Dedicated energy crops, such as switchgrass, are also promising cellulose sources that can be sustainably produced in many regions of the United States.

The world's first commercial tidal stream generator – SeaGen – in Strangford Lough. The strong wake shows the power in the tidal current.

In terms of ocean energy, another third-generation technology, Portugal has the world's first commercial wave farm, the Aguçadora Wave Park, under construction in 2007. The farm will initially use three Pelamis P-750 machines generating 2.25 MW. and costs are put at 8.5 million euro. Subject to successful operation, a further 70 million euro is likely to be invested before 2009 on a further 28 machines to generate 525 MW. Funding for a wave farm in Scotland was announced in February, 2007 by the Scottish Executive, at a cost of over 4 million pounds, as part of a £13 million funding packages for ocean power in Scotland. The farm will be the world's largest with a capacity of 3 MW generated by four Pelamis machines.

In 2007, the world's first turbine to create commercial amounts of energy using tidal power was installed in the narrows of Strangford Lough in Ireland. The 1.2 MW underwater tidal electricity generator takes advantage of the fast tidal flow in the lough which can be up to 4m/s. Although the generator is powerful enough to power up to a thousand homes, the turbine has a minimal environmental impact, as it is almost entirely submerged, and the rotors turn slowly enough that they pose no danger to wildlife.

Solar power panels that use nanotechnology, which can create circuits out of individual silicon molecules, may cost half as much as traditional photovoltaic cells, according to executives and investors involved in developing the products. Nanosolar has secured more than $100 million from investors to build a factory for nanotechnology thin-film solar panels. The company's plant has a planned production capacity of 430 megawatts peak power of solar cells per year. Commercial production started and first panels have been shipped to customers in late 2007.

Large national and regional research projects on artificial photosynthesis are designing nanotechnology-based systems that use solar energy to split water into hydrogen fuel. and a proposal has been made for a Global Artificial Photosynthesis project In 2011, researchers at the Massachusetts Institute of Technology (MIT) developed what they are calling an "Artificial Leaf", which is capable of splitting water into hydrogen and oxygen directly from solar power when dropped into a glass of water. One side of the "Artificial Leaf" produces bubbles of hydrogen, while the other side produces bubbles of oxygen.

Most current solar power plants are made from an array of similar units where each unit is continuously adjusted, e.g., with some step motors, so that the light converter stays in focus of the sun light. The cost of focusing light on converters such as high-power solar panels, Stirling engine, etc. can be dramatically decreased with a simple and efficient rope mechanics. In this technique many units are connected with a network of ropes so that pulling two or three ropes is sufficient to keep all light converters simultaneously in focus as the direction of the sun changes.

Japan and China have national programs aimed at commercial scale Space-Based Solar Power (SBSP). The China Academy of Space Technology (CAST) won the 2015 International SunSat Design Competition with this video of their Multi-Rotary Joint design. Proponents of SBSP claim that Space-Based Solar Power would be clean, constant, and global, and could scale to meet all planetary energy demand. A recent multi-agency industry proposal (echoing the 2008 Pentagon recommendation) won the SECDEF/SECSTATE/USAID Director D3 (Diplomacy, Development, Defense) Innovation Challenge.

Enabling Technologies for Renewable Energy

Heat pumps and Thermal energy storage are classes of technologies that can enable the utilization of renewable energy sources that would otherwise be inaccessible due to a temperature that is too low for utilization or a time lag between when the energy is available and when it is needed. While enhancing the temperature of available renewable thermal energy, heat pumps have the additional property of leveraging electrical power (or in some cases mechanical or thermal power) by using it to extract additional energy from a low quality source (such as seawater, lake water, the ground, the air, or waste heat from a process).

Thermal storage technologies allow heat or cold to be stored for periods of time ranging from hours or overnight to interseasonal, and can involve storage of sensible energy (i.e. by changing the temperature of a medium) or latent energy (i.e. through phase changes of a medium, such between water and slush or ice). Short-term thermal storages can be used for peak-shaving in district heating or electrical distribution systems. Kinds of renewable or alternative energy sources that can be enabled include natural energy (e.g. collected via solar-thermal collectors, or dry cooling towers used to collect winter's cold), waste energy (e.g. from HVAC equipment, industrial processes or power plants), or surplus energy (e.g. as seasonally from hydropower projects or intermittently from wind farms). The Drake Landing Solar Community (Alberta, Canada) is illustrative. borehole thermal energy storage allows the community to get 97% of its year-round heat from solar collectors on the garage roofs, which most of the heat collected in summer. Types of storages for sensible energy include insulated tanks, borehole clusters in substrates ranging from gravel to bedrock, deep aquifers, or shallow lined pits that are insulated on top. Some types of storage are capable of storing heat or cold between opposing seasons (particularly if very large), and some storage applications require inclusion of a heat pump. Latent heat is typically stored in ice tanks or what are called phase-change materials (PCMs).

Energy Efficiency

Moving towards energy sustainability will require changes not only in the way energy is supplied, but in the way it is used, and reducing the amount of energy required to deliver various goods or services is essential. Opportunities for improvement on the demand side of the energy equation are as rich and diverse as those on the supply side, and often offer significant economic benefits.

Renewable energy and energy efficiency are sometimes said to be the "twin pillars" of sustainable energy policy. Both resources must be developed in order to stabilize and reduce carbon dioxide emissions. Efficiency slows down energy demand growth so that rising clean energy supplies can make deep cuts in fossil fuel use. If energy use grows too fast, renewable energy development will chase a receding target. A recent historical analysis has demonstrated that the rate of energy efficiency improvements has generally been outpaced by the rate of growth in energy demand, which is due to continuing economic and population growth. As a result, despite energy efficiency gains, total energy use and related carbon emissions have continued to increase. Thus, given the thermodynamic and practical limits of energy efficiency improvements, slowing the growth in energy demand is essential. However, unless clean energy supplies come online rapidly, slowing demand growth will only begin to reduce total emissions; reducing the carbon content of energy sources is also needed. Any serious vision of a sustainable energy economy thus requires commitments to both renewables and efficiency.

Renewable energy (and energy efficiency) are no longer niche sectors that are promoted only by governments and environmentalists. The increased levels of investment and the fact that much of the capital is coming from more conventional financial actors suggest that sustainable energy options are now becoming mainstream. An example of this would be The Alliance to Save Energy's Project with Stahl Consolidated Manufacturing, (Huntsville, Alabama, USA) (StahlCon 7), a patented generator shaft designed to reduce emissions within existing power generating systems, granted publishing rights to the Alliance in 2007.

Climate change concerns coupled with high oil prices and increasing government support are driving increasing rates of investment in the sustainable energy industries, according to a trend analysis from the United Nations Environment Programme. According to UNEP, global investment in sustainable energy in 2007 was higher than previous levels, with $148 billion of new money raised in 2007, an increase of 60% over 2006. Total financial transactions in sustainable energy, including acquisition activity, was $204 billion.

Investment flows in 2007 broadened and diversified, making the overall picture one of greater breadth and depth of sustainable energy use. The mainstream capital markets are "now fully receptive to sustainable energy companies, supported by a surge in funds destined for clean energy investment".

Smart-grid Technology

Smart grid refers to a class of technology people are using to bring utility electricity delivery systems into the 21st century, using computer-based remote control and automation. These systems are made possible by two-way communication technology and computer processing that has been used for decades in other industries. They are beginning to be used on electricity networks, from the power plants and wind farms all the way to the consumers of electricity in homes and businesses. They offer many benefits to utilities and consumers—mostly seen in big improvements in energy efficiency on the electricity grid and in the energy users' homes and offices.

Green Energy and Green Power

A solar trough array is an example of green energy

Public seat with integrated solar panel in Singapore. Anyone can sit and plug in their mobile phone for free charging

Green energy includes natural energetic processes that can be harnessed with little pollution. Green power is electricity generated from renewable energy sources.

Anaerobic digestion, geothermal power, wind power, small-scale hydropower, solar energy, biomass power, tidal power, wave power, and some forms of nuclear power (ones which are able to "burn" nuclear waste through a process known as nuclear transmutation, such as an Integral Fast Reactor, and therefore belong in the "Green Energy" category). Some definitions may also include power derived from the incineration of waste.

Some people, including Greenpeace founder and first member Patrick Moore, George Monbiot, Bill Gates and James Lovelock have specifically classified nuclear power as green energy. Others, including Greenpeace's Phil Radford disagree, claiming that the problems associated with radioactive waste and the risk of nuclear accidents (such as the Chernobyl disaster) pose an unacceptable risk to the environment and to humanity. However, newer nuclear reactor designs are capable of utilizing what is now deemed "nuclear waste" until it is no longer (or dramatically less) dangerous, and have design features that greatly minimize the possibility of a nuclear accident. These designs have yet to be proven.

Some have argued that although green energy is a commendable effort in solving the world's increasing energy consumption, it must be accompanied by a cultural change that encourages the decrease of the world's appetite for energy.

In several countries with common carrier arrangements, electricity retailing arrangements make it possible for consumers to purchase green electricity (renewable electricity) from either their utility or a green power provider.

When energy is purchased from the electricity network, the power reaching the consumer will not necessarily be generated from green energy sources. The local utility company, electric company, or state power pool buys their electricity from electricity producers who may be generating from fossil fuel, nuclear or renewable energy sources. In many countries green energy currently provides a very small amount of electricity, generally contributing less than 2 to 5% to the overall pool. In some U.S. states, local governments have formed regional power purchasing pools using Community Choice Aggregation and Solar Bonds to achieve a 51% renewable mix or higher, such as in the City of San Francisco.

By participating in a green energy program a consumer may be having an effect on the energy sources used and ultimately might be helping to promote and expand the use of green energy. They are also making a statement to policy makers that they are willing to pay a price premium to support renewable energy. Green energy consumers either obligate the utility companies to increase the amount of green energy that they purchase from the pool (so decreasing the amount of non-green energy they purchase), or directly fund the green energy through a green power provider. If insufficient green energy sources are available, the utility must develop new ones or contract with a third party energy supplier to provide green energy, causing more to be built. However, there is no way the consumer can check whether or not the electricity bought is "green" or otherwise.

In some countries such as the Netherlands, electricity companies guarantee to buy an equal amount of 'green power' as is being used by their green power customers. The Dutch government exempts green power from pollution taxes, which means green power is hardly any more expensive than other power.

In the United States, one of the main problems with purchasing green energy through the electrical grid is the current centralized infrastructure that supplies the consumer's electricity. This infrastructure has led to increasingly frequent brown outs and black outs, high CO_2 emissions, higher energy costs, and power quality issues. An additional $450 billion will be invested to expand this fledgling system over the next 20 years to meet increasing demand. In addition, this centralized system is now being further over-taxed with the incorporation of renewable energies such as wind, solar, and geothermal energies. Renewable resources, due to the amount of space they require, are often located in remote areas where there is a lower energy demand. The current infrastructure would make transporting this energy to high demand areas, such as urban centers, highly inefficient and in some cases impossible. In addition, despite the amount of renewable energy produced or the economic viability of such technologies only about 20 percent will be able to be incorporated into the grid. To have a more sustainable energy profile, the United States must move towards implementing changes to the electrical grid that will accommodate a mixed-fuel economy.

However, several initiatives are being proposed to mitigate these distribution problems. First and foremost, the most effective way to reduce USA's CO_2 emissions and slow global warming is through conservation efforts. Opponents of the current US electrical grid have also advocated for decentralizing the grid. This system would increase efficiency by reducing the amount of energy lost in transmission. It would also be economically viable as it would reduce the amount of power lines that will need to be constructed in the future to keep up with demand. Merging heat and power in this system would create added benefits and help to increase its efficiency by up to 80-90%. This is a significant increase from the current fossil fuel plants which only have an efficiency of 34%.

A more recent concept for improving our electrical grid is to beam microwaves from Earth-orbiting satellites or the moon to directly when and where there is demand. The power would be generated from solar energy captured on the lunar surface In this system, the receivers would be "broad, translucent tent-like structures that would receive microwaves and convert them to electricity". NASA said in 2000 that the technology was worth pursuing but it is still too soon to say if the technology will be cost-effective.

The World Wide Fund for Nature and several green electricity labelling organizations created the (now defunct) Eugene Green Energy Standard under which the national green electricity certification schemes could be accredited to ensure that the purchase of green energy leads to the provision of additional new green energy resources.

Local Green Energy Systems

A small Quietrevolution QR5 Gorlov type vertical axis wind turbine in Bristol, England. Measuring 3 m in diameter and 5 m high, it has a nameplate rating of 6.5 kW to the grid.

Those not satisfied with the third-party grid approach to green energy via the power grid can install their own locally based renewable energy system. Renewable energy electrical systems from solar to wind to even local hydro-power in some cases, are some of the many types of renewable energy systems available locally. Additionally, for those interested in heating and cooling their dwelling via renewable energy, geothermal heat pump systems that tap the constant temperature of the earth, which is around 7 to 15 degrees Celsius a few feet underground and increases dramatically at greater depths, are an option over conventional natural gas and petroleum-fueled heat approaches. Also, in geographic locations where the Earth's Crust is especially thin, or near volcanoes (as is the case in Iceland) there exists the potential to generate even more electricity than would be possible at other sites, thanks to a more significant temperature gradient at these locales.

The advantage of this approach in the United States is that many states offer incentives to offset the cost of installation of a renewable energy system. In California, Massachusetts and several other U.S. states, a new approach to community energy supply called Community Choice Aggregation has provided communities with the means to solicit a competitive electricity supplier and use municipal revenue bonds to finance development of local green energy resources. Individuals are usually assured that the electricity they are using is actually produced from a green energy source that they control. Once the system is paid for, the owner of a renewable energy system will be producing their own renewable electricity for essentially no cost and can sell the excess to the local utility at a profit.

Using Green Energy

A 01 KiloWatt Micro Windmill for Domestic Usage

Renewable energy, after its generation, needs to be stored in a medium for use with autonomous devices as well as vehicles. Also, to provide household electricity in remote areas (that is areas which are not connected to the mains electricity grid), energy storage is required for use with renewable energy. Energy generation and consumption systems used in the latter case are usually stand-alone power systems.

Some examples are:

- energy carriers as hydrogen, liquid nitrogen, compressed air, oxyhydrogen, batteries, to power vehicles.
- flywheel energy storage, pumped-storage hydroelectricity is more usable in stationary applications (e.g. to power homes and offices). In household power systems, conversion of energy can also be done to reduce smell. For example, organic matter such as cow dung and spoilable organic matter can be converted to biochar. To eliminate emissions, carbon capture and storage is then used.

Usually however, renewable energy is derived from the mains electricity grid. This means that energy storage is mostly not used, as the mains electricity grid is organised to produce the exact amount of energy being consumed at that particular moment. Energy production on the mains electricity grid is always set up as a combination of (large-scale) renewable energy plants, as well as other power plants as fossil-fuel power plants and nuclear power. This combination however, which is essential for this type of energy supply (as e.g. wind turbines, solar power plants etc.) can only produce when the wind blows and the sun shines. This is also one of the main drawbacks of the system as fossil fuel powerplants are polluting and are a main cause of global warming (nuclear power being an exception). Although fossil fuel power plants too can be made emissionless (through carbon capture and storage), as well as renewable (if the plants are converted to e.g. biomass) the best solution is still to phase out the latter power plants over time. Nuclear power plants too can be more or less eliminated from their problem of nuclear waste through the use of nuclear reprocessing and newer plants as fast breeder and nuclear fusion plants.

Renewable energy power plants do provide a steady flow of energy. For example, hydropower plants, ocean thermal plants, osmotic power plants all provide power at a regulated pace, and are thus available power sources at any given moment (even at night, windstill moments etc.). At present however, the number of steady-flow renewable energy plants alone is still too small to meet energy demands at the times of the day when the irregular producing renewable energy plants cannot produce power.

Besides the greening of fossil fuel and nuclear power plants, another option is the distribution and immediate use of power from solely renewable sources. In this set-up energy storage is again not necessary. For example, TREC has proposed to distribute solar power from the Sahara to Europe. Europe can distribute wind and ocean power to the Sahara and other countries. In this way, power is produced at any given time as at any point of the planet as the sun or the wind is up or ocean waves and currents are stirring. This option however is probably not possible in the short-term, as fossil fuel and nuclear power are still the main sources of energy on the mains electricity net and replacing them will not be possible overnight.

Several large-scale energy storage suggestions for the grid have been done. Worldwide there is over 100 GW of Pumped-storage hydroelectricity. This improves efficiency and decreases energy losses but a conversion to an energy storing mains electricity grid is a very costly solution. Some costs could potentially be reduced by making use of energy storage equipment the consumer buys and not the state. An example is batteries in electric cars that would double as an energy buffer for the electricity grid. However besides the cost, setting-up such a system would still be a very complicated and difficult procedure. Also, energy storage apparatus' as car batteries are also built with materials that pose a threat to the environment (e.g. Lithium). The combined production of batteries for such a large part of the population would still have environmental concerns. Besides car batteries however, other Grid energy storage projects make use of less polluting energy carriers (e.g. compressed air tanks and flywheel energy storage).

Carbon-neutral and Negative Fuels

A carbon-neutral fuel is a synthetic fuel – such as methane, gasoline, diesel fuel or jet fuel – produced from renewable or nuclear energy used to hydrogenate waste carbon dioxide recycled from power plant flue-gas emissions, recovered from automotive exhaust gas, or derived from carbonic acid in seawater. Such fuels are carbon-neutral because they do not result in a net increase in atmospheric greenhouse gases. To the extent that carbon-neutral fuels displace fossil fuels, or if they are produced from waste carbon or seawater carbonic acid, and their combustion is subject to carbon capture at the flue or exhaust pipe, they result in negative carbon dioxide emission and net carbon dioxide removal from the atmosphere, and thus constitute a form of greenhouse gas remediation. Such fuels are produced by the electrolysis of water to make hydrogen used in turn in the Sabatier reaction to produce methane which may then be stored to be burned later in power plants as synthetic natural gas, transported by pipeline, truck, or tanker ship, or be used in gas to liquids processes such as the Fischer–Tropsch process to make traditional transportation or heating fuels.

Green Energy and Labeling By Region

European Union

Directive 2004/8/EC of the European Parliament and of the Council of 11 February 2004 on the promotion of cogeneration based on a useful heat demand in the internal energy market includes the article 5 (*Guarantee of origin of electricity* from high-efficiency cogeneration).

European environmental NGOs have launched an ecolabel for green power. The ecolabel is called EKOenergy. It sets criteria for sustainability, additionality, consumer information and tracking. Only part of electricity produced by renewables fulfills the EKOenergy criteria.

A Green Energy Supply Certification Scheme was launched in the United Kingdom in February 2010. This implements guidelines from the Energy Regulator, Ofgem, and sets requirements on transparency, the matching of sales by renewable energy supplies, and additionality.

United States

The United States Department of Energy (DOE), the Environmental Protection Agency (EPA), and the Center for Resource Solutions (CRS) recognizes the voluntary purchase of electricity from renewable energy sources (also called renewable electricity or green electricity) as green power.

The most popular way to purchase renewable energy as revealed by NREL data is through purchasing Renewable Energy Certificates (RECs). According to a Natural Marketing Institute (NMI) survey 55 percent of American consumers want companies to increase their use of renewable energy.

DOE selected six companies for its 2007 Green Power Supplier Awards, including Constellation NewEnergy; 3Degrees; Sterling Planet; SunEdison; Pacific Power and Rocky Mountain Power; and Silicon Valley Power. The combined green power provided by those six winners equals more than 5 billion kilowatt-hours per year, which is enough to power nearly 465,000 average U.S. households. In 2014, Arcadia Power made RECS available to homes and businesses in all 50 states, allowing consumers to use "100% green power" as defined by the EPA's Green Power Partnership.

The U.S. Environmental Protection Agency (USEPA) Green Power Partnership is a voluntary program that supports the organizational procurement of renewable electricity by offering expert advice, technical support, tools and resources. This can help organizations lower the transaction costs of buying renewable power, reduce carbon footprint, and communicate its leadership to key stakeholders.

Throughout the country, more than half of all U.S. electricity customers now have an option to purchase some type of green power product from a retail electricity provider. Roughly one-quarter of the nation's utilities offer green power programs to customers, and voluntary retail sales of renewable energy in the United States totaled more than 12 billion kilowatt-hours in 2006, a 40% increase over the previous year.

Sustainable Energy Research

There are numerous organizations within the academic, federal, and commercial sectors conducting large scale advanced research in the field of sustainable energy. This research spans several areas of focus across the sustainable energy spectrum. Most of the research is targeted at improving efficiency and increasing overall energy yields. Multiple federally supported research organizations have focused on sustainable energy in recent years. Two of the most prominent of these labs are Sandia National Laboratories and the National Renewable Energy Laboratory (NREL), both of which are funded by the United States Department of Energy and supported by various corporate partners. Sandia has a total budget of $2.4 billion while NREL has a budget of $375 million.

Scientific production towards sustainable energy systems is rising exponentially, growing from about 500 English journal papers only about renewable energy in 1992 to almost 9,000 papers in 2011.

Solar

The primary obstacle that is preventing the large scale implementation of solar powered energy generation is the inefficiency of current solar technology. Currently, photovoltaic (PV) panels only have the ability to convert around 16% of the sunlight that hits them into electricity. At this rate, many experts believe that solar energy is not efficient enough to be economically sustainable given the cost to produce the panels themselves. Both Sandia National Laboratories and the National Renewable Energy Laboratory (NREL), have heavily funded solar research programs. The NREL solar program has a budget of around $75 million and develops research projects in the areas of photovoltaic (PV) technology, solar thermal energy, and solar radiation. The budget for Sandia's solar division is unknown, however it accounts for a significant percentage of the laboratory's $2.4 billion budget. Several academic programs have focused on solar research in recent years. The Solar Energy Research Center (SERC) at University of North Carolina (UNC) has the sole purpose of developing cost effective solar technology. In 2008, researchers at Massachusetts Institute of Technology (MIT) developed a method to store solar energy by using it to produce hydrogen fuel from water. Such research is targeted at addressing the obstacle that solar development faces of storing energy for use during nighttime hours when the sun is not shining. In February 2012, North Carolina-based Semprius Inc., a solar development company backed by German corporation Siemens, announced that they had developed the world's most efficient solar panel. The company claims that the prototype converts 33.9% of the sunlight that hits it to electricity, more than double the previous high-end conversion rate. Major projects on artificial photosynthesis or solar fuels are also under way in many developed nations.

Space-Based Solar Power

Space-Based Solar Power Satellites seek to overcome the problems of storage and provide civilization-scale power that is clean, constant, and global. Japan and China have active national programs aimed at commercial scale Space-Based Solar Power (SBSP), and both nation's hope to orbit demonstrations in the 2030s. The China Academy of Space Technology (CAST) won the 2015 International SunSat Design Competition with this video of their Multi-Rotary Joint design. Proponents of SBSP claim that Space-Based Solar Power would be clean, constant, and global, and could scale to meet all planetary energy demand. A recent multi-agency industry proposal (echoing the 2008 Pentagon recommendation) won the SECDEF/SECSTATE/USAID Director D3 (Diplomacy, Development, Defense) Innovation Challenge with the following pitch and vision video. Northrop Grumman is funding CALTECH with $17.5 million for an ultra lightweight design. Keith Henson recently posted a video of a "bootstrapping" approach.

Wind

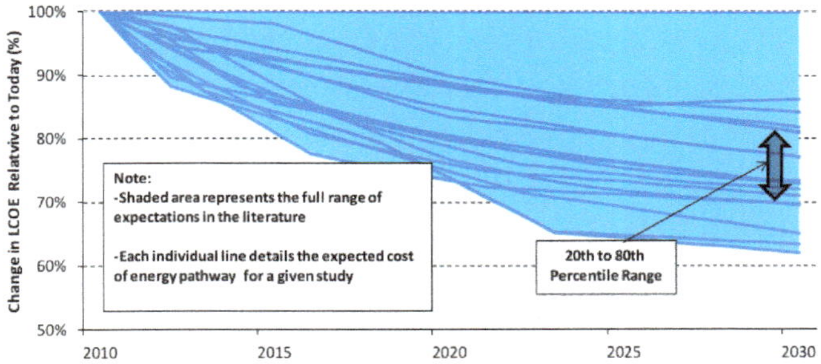

The National Renewable Energy Laboratory projects that the levelized cost of wind power in the U.S. will decline about 25% from 2012 to 2030.

Wind energy research dates back several decades to the 1970s when NASA developed an analytical model to predict wind turbine power generation during high winds. Today, both Sandia National Laboratories and National Renewable Energy Laboratory have programs dedicated to wind research. Sandia's laboratory focuses on the advancement of materials, aerodynamics, and sensors. The NREL wind projects are centered on improving wind plant power production, reducing their capital costs, and making wind energy more cost effective overall. The Field Laboratory for Optimized Wind Energy (FLOWE) at Caltech was established to research renewable approaches to wind energy farming technology practices that have the potential to reduce the cost, size, and environmental impact of wind energy production. The president of Sky WindPower Corporation thinks that wind turbines will be able to produce electricity at a cent/kWh at an average which in comparison to coal-generated electricity is a fractional of the cost.

Bangui Wind Farm in the Philippines.

A wind farm is a group of wind turbines in the same location used to produce electric power. A large wind farm may consist of several hundred individual wind turbines, and cover an extended area of hundreds of square miles, but the land between the turbines may be used for agricultural or other purposes. A wind farm may also be located offshore.

Many of the largest operational onshore wind farms are located in the USA and China. The Gansu Wind Farm in China has over 5,000 MW installed with a goal of 20,000 MW by 2020. China has several other "wind power bases" of similar size. The Alta Wind Energy Center in California is the largest onshore wind farm outside of China, with a capacity of 1020 MW of power. Europe leads in the use of wind power with almost 66 GW, about 66 percent of the total globally, with Denmark in the lead according to the countries installed per-capita capacity. As of February 2012, the Walney Wind Farm in United Kingdom is the largest offshore wind farm in the world at 367 MW, followed by Thanet Wind Farm (300 MW), also in the UK.

There are many large wind farms under construction and these include BARD Offshore 1 (400 MW), Clyde Wind Farm (350 MW), Greater Gabbard wind farm (500 MW), Lincs Wind Farm (270 MW), London Array (1000 MW), Lower Snake River Wind Project (343 MW), Macarthur Wind Farm (420 MW), Shepherds Flat Wind Farm (845 MW), and Sheringham Shoal (317 MW).

Wind power has expanded quickly, its share of worldwide electricity usage at the end of 2014 was 3.1%.

Carbon-neutral and Negative Fuels

Carbon-neutral fuels are synthetic fuels (including methane, gasoline, diesel fuel, jet fuel or ammonia) produced by hydrogenating waste carbon dioxide recycled from power plant flue-gas emissions, recovered from automotive exhaust gas, or derived from carbonic acid in seawater. Commercial fuel synthesis companies suggest they can produce synthetic fuels for less than petroleum fuels when oil costs more than $55 per barrel. Renewable methanol (RM) is a fuel produced from hydrogen and carbon dioxide by catalytic hydrogenation where the hydrogen has been obtained from water electrolysis. It can be blended into transportation fuel or processed as a chemical feedstock.

The George Olah carbon dioxide recycling plant operated by Carbon Recycling International in Grindavík, Iceland has been producing 2 million liters of methanol transportation fuel per year from flue exhaust of the Svartsengi Power Station since 2011. It has the capacity to produce 5 million liters per year. A 250 kilowatt methane synthesis plant was constructed by the Center for Solar Energy and Hydrogen Research (ZSW) at Baden-Württemberg and the Fraunhofer Society in Germany and began operating in 2010. It is being upgraded to 10 megawatts, scheduled for completion in autumn, 2012. Audi has constructed a carbon-neutral liquefied natural gas (LNG) plant in Werlte, Germany. The plant is intended to produce transportation fuel to offset LNG used in their A3 Sportback g-tron automobiles, and can keep 2,800 metric tons of CO_2 out of the environment per year at its initial capacity. Other commercial developments are taking place in Columbia, South Carolina, Camarillo, California, and Darlington, England.

Such fuels are considered carbon-neutral because they do not result in a net increase in atmospheric greenhouse gases. To the extent that synthetic fuels displace fossil fuels, or if they are produced from waste carbon or seawater carbonic acid, and their combustion is subject to carbon capture at the flue or exhaust pipe, they result in negative carbon dioxide emission and net carbon dioxide removal from the atmosphere, and thus constitute a form of greenhouse gas remediation.

Such renewable fuels alleviate the costs and dependency issues of imported fossil fuels without requiring either electrification of the vehicle fleet or conversion to hydrogen or other fuels, enabling continued compatible and affordable vehicles. Carbon-neutral fuels offer relatively low cost energy storage, alleviating the problems of wind and solar intermittency, and they enable distribution of wind, water, and solar power through existing natural gas pipelines.

Nighttime wind power is considered the most economical form of electrical power with which to synthesize fuel, because the load curve for electricity peaks sharply during the warmest hours of the day, but wind tends to blow slightly more at night than during the day, so, the price of nighttime wind power is often much less expensive than any alternative. Germany has built a 250 kilowatt synthetic methane plant which they are scaling up to 10 megawatts.

Biomass

Sugarcane plantation to produce ethanol in Brazil

A CHP power station using wood to provide electricity to over 30.000 households in France

Biomass is biological material derived from living, or recently living organisms. It most often refers to plants or plant-derived materials which are specifically called lignocellulosic biomass. As an energy source, biomass can either be used directly via combustion to produce heat, or indirectly after converting it to various forms of biofuel. Conversion of biomass to biofuel can be achieved by different methods which are broadly classified into: *thermal*, *chemical*, and *biochemical* methods. Wood remains the largest biomass energy source today; examples include forest residues – such as dead trees, branches and tree stumps –, yard clippings, wood chips and even municipal solid waste. In the second sense, biomass includes plant or animal matter that can be converted into fibers or other industrial chemicals, including biofuels. Industrial biomass can be grown from numerous types of plants, including miscanthus, switchgrass, hemp, corn, poplar, willow, sorghum, sugarcane, bamboo, and a variety of tree species, ranging from eucalyptus to oil palm (palm oil).

Biomass, biogas and biofuels are burned to produce heat/power and in doing so harm the environment. Pollutants such as sulphurous oxides (SO_x), nitrous oxides (NO_x), and particulate matter (PM) are produced from this combustion; the World Health Organisation estimates that 7 million premature deaths are caused each year by air pollution. Biomass combustion is a major contributor.

Ethanol Biofuels

As the primary source of biofuel in North America, many organizations are conducting research in the area of ethanol production. On the Federal level, the USDA conducts a large amount of research regarding ethanol production in the United States. Much of this research is targeted towards the effect of ethanol production on domestic food markets. The National Renewable Energy Laboratory has conducted various ethanol research projects, mainly in the area of cellulosic ethanol. Cellulosic ethanol has many benefits over traditional corn based-ethanol. It does not take away or directly conflict with the food supply because it is produced from wood, grasses, or non-edible parts of plants. Moreover, some studies have shown cellulosic ethanol to be more cost effective and economically sustainable than corn-based ethanol. Even if we used all the corn crop that we have in the United States and converted it into ethanol it would only produce enough fuel to serve 13 percent of the United States total gasoline consumption. Sandia National Laboratories conducts in-house cellulosic ethanol research and is also a member of the Joint BioEnergy Institute (JBEI), a research institute founded by the United States Department of Energy with the goal of developing cellulosic biofuels.

Other Biofuels

From 1978 to 1996, the National Renewable Energy Laboratory experimented with producing algae fuel in the "Aquatic Species Program." A self-published article by Michael Briggs, at the University of New Hampshire Biofuels Group, offers estimates for the realistic replacement of all motor vehicle fuel with biofuels by utilizing algae that have a natural oil content greater than 50%, which Briggs suggests can be grown on algae ponds at wastewater treatment plants. This oil-rich algae can then be extracted from the system and processed into biofuels, with the dried remainder further reprocessed to create ethanol. The production of algae to harvest oil for biofuels has not yet been undertaken on a commercial scale, but feasibility studies have been conducted to arrive at the above yield estimate. During the biofuel production process algae actually consumes the carbon dioxide in the air and turns it into oxygen through photosynthesis. In addition to its projected high yield, algaculture— unlike food crop-based biofuels — does not entail a decrease in food production, since it requires neither farmland nor fresh water. Many companies are pursuing algae bio-reactors for various purposes, including scaling up biofuels production to commercial levels.

Several groups in various sectors are conducting research on Jatropha curcas, a poisonous shrub-like tree that produces seeds considered by many to be a viable source of biofuels feedstock oil. Much of this research focuses on improving the overall per acre oil yield of Jatropha through advancements in genetics, soil science, and horticultural practices. SG Biofuels, a San Diego-based Jatropha developer, has used molecular breeding and biotechnology to produce elite hybrid seeds of Jatropha that show significant yield improvements over first generation varieties. The Center for Sustainable Energy Farming (CfSEF) is a Los Angeles-based non-profit research organization dedicated to Jatropha research in the areas of plant science, agronomy, and horticulture. Successful exploration of these disciplines is projected to increase Jatropha farm production yields by 200-300% in the next ten years.

Geothermal

Geothermal energy is produced by tapping into the thermal energy created and stored within the earth. It arises from the radioactive decay of an isotope of potassium and other elements found in the Earth's crust. Geothermal energy can be obtained by drilling into the ground, very similar to oil exploration, and then it is carried by a heat-transfer fluid (e.g. water, brine or steam). Geothermal systems that are mainly dominated by water have the potential to provide greater benefits to the system and will generate more power. Within these liquid-dominated systems, there are possible concerns of subsidence and contamination of ground-water resources. Therefore, protection of ground-water resources is necessary in these systems. This means that careful reservoir production and engineering is necessary in liquid-dominated geothermal reservoir systems. Geothermal energy is considered sustainable because that thermal energy is constantly replenished. However, the science of geothermal energy generation is still young and developing economic viability. Several entities, such as the National Renewable Energy Laboratory and Sandia National Laboratories are conducting research toward the goal of establishing a proven science around geothermal energy. The International Centre for Geothermal Research (IGC), a German geosciences research organization, is largely focused on geothermal energy development research.

Hydrogen

Over $1 billion of federal money has been spent on the research and development of hydrogen and a medium for energy storage in the United States. Both the National Renewable Energy Laboratory and Sandia National Laboratories have departments dedicated to hydrogen research. Hydrogen is useful for energy storage and for use in airplanes, but is not practical for automobile use, as it is not very efficient, compared to using a battery — for the same cost a person can travel three times as far using a battery.

Thorium

There are potentially two sources of nuclear power. Fission is used in all current nuclear power plants. Fusion is the reaction that exists in stars, including the sun, and remains impractical for use on Earth, as fusion reactors are not yet available. However nuclear power is controversial politically and scientifically due to concerns about radioactive waste disposal, safety, the risks of a severe accident, and technical and economical problems in dismantling of old power plants.

Thorium is a fissionable material used in thorium-based nuclear power. The thorium fuel cycle claims several potential advantages over a uranium fuel cycle, including greater abundance, superior physical and nuclear properties, better resistance to nuclear weapons proliferation and reduced plutonium and actinide production. Therefore, it is sometimes referred as sustainable.

Clean Energy Investments

2010 was a record year for green energy investments. According to a report from Bloomberg New Energy Finance, nearly US $243 billion was invested in wind farms, solar power, electric cars, and other alternative technologies worldwide, representing a 30 percent increase from 2009 and nearly five times the money invested in 2004. China had $51.1 billion investment in clean energy projects in 2010, by far the largest figure for any country.

Within emerging economies, Brazil comes second to China in terms of clean energy investments. Supported by strong energy policies, Brazil has one of the world's highest biomass and small-hydro power capacities and is poised for significant growth in wind energy investment. The cumulative investment potential in Brazil from 2010 to 2020 is projected as $67 billion.

India is another rising clean energy leader. While India ranked the 10th in private clean energy investments among G-20 members in 2009, over the next 10 years it is expected to rise to the third position, with annual clean energy investment under current policies forecast to grow by 369 percent between 2010 and 2020.

It is clear that the center of growth has started to shift to the developing economies and they may lead the world in the new wave of clean energy investments.

Around the world many sub-national governments - regions, states and provinces - have aggressively pursued sustainable energy investments. In the United States, California's leadership in renewable energy was recognised by The Climate Group when it awarded former Governor Arnold Schwarzenegger its inaugural award for international climate leadership in Copenhagen in 2009. In Australia, the state of South Australia - under the leadership of former Premier Mike Rann - has led the way with wind power comprising 26% of its electricity generation by the end of 2011, edging out coal fired generation for the first time. South Australia also has had the highest take-up per capita of household solar panels in Australia following the Rann Government's introduction of solar feed-in laws and educative campaign involving the installation of solar photovoltaic installations on the roofs of prominent public buildings, including the parliament, museum, airport and Adelaide Showgrounds pavilion and schools. Rann, Australia's first climate change minister, passed legislation in 2006 setting targets for renewable energy and emissions cuts, the first legislation in Australia to do so.

Also, in the European Union there is a clear trend of promoting policies encouraging investments and financing for sustainable energy in terms of energy efficiency, innovation in energy exploitation and development of renewable resources, with increased consideration of environmental aspects and sustainability.

Renewable Energy

Wind, solar, and hydroelectricity are three emerging renewable sources of energy

Renewable energy is generally defined as energy that is collected from resources which are naturally replenished on a human timescale, such as sunlight, wind, rain, tides, waves, and geothermal heat. Renewable energy often provides energy in four important areas: electricity generation, air and water heating/cooling, transportation, and rural (off-grid) energy services.

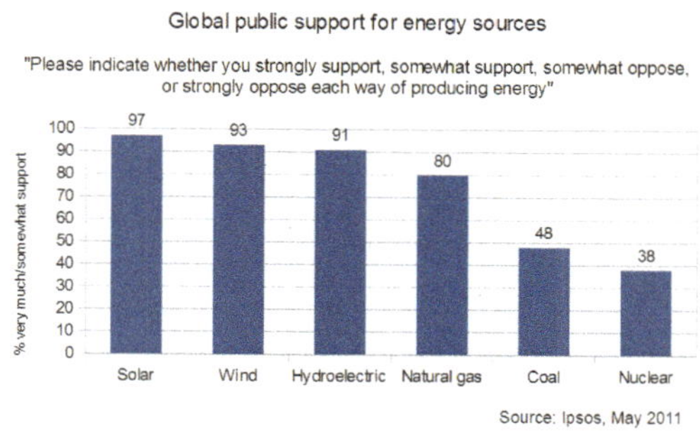

Global public support for different energy sources (2011) based on a poll by Ipsos Global @dvisor

Based on REN21's 2016 report, renewables contributed 19.2% to humans' global energy consumption and 23.7% to their generation of electricity in 2014 and 2015, respectively. This energy consumption is divided as 8.9% coming from traditional biomass, 4.2% as heat energy (modern biomass, geothermal and solar heat), 3.9% hydro electricity and 2.2% is electricity from wind, solar, geothermal, and biomass. Worldwide investments in renewable technologies amounted to more than US$286 billion in 2015, with countries like China and the United States heavily investing in wind, hydro, solar and biofuels. Globally, there are an estimated 7.7 million jobs associated with the renewable energy industries, with solar photovoltaics being the largest renewable employer.

Renewable energy resources exist over wide geographical areas, in contrast to other energy sources, which are concentrated in a limited number of countries. Rapid deployment of renewable energy and energy efficiency is resulting in significant energy security, climate change mitigation, and economic benefits. The results of a recent review of the literature concluded that as greenhouse gas (GHG) emitters begin to be held liable for damages resulting from GHG emissions resulting in climate change, a high value for liability mitigation would provide powerful incentives for deployment of renewable energy technologies. In international public opinion surveys there is strong support for promoting renewable sources such as solar power and wind power. At the national level, at least 30 nations around the world already have renewable energy contributing more than 20 percent of energy supply. National renewable energy markets are projected to continue to grow strongly in the coming decade and beyond. Some places and at least two countries, Iceland and Norway generate all their electricity using renewable energy already, and many other countries have the set a goal to reach 100% renewable energy in the future. For example, in Denmark the government decided to switch the total energy supply (electricity, mobility and heating/cooling) to 100% renewable energy by 2050.

While many renewable energy projects are large-scale, renewable technologies are also suited to rural and remote areas and developing countries, where energy is often crucial in human development. United Nations' Secretary-General Ban Ki-moon has said that renewable energy has the ability to lift the poorest nations to new levels of prosperity. As most of renewables provide electricity, renewable energy deployment is often applied in conjunction with further electrification, which has several benefits: For example, electricity can be converted to heat without losses and even reach higher temperatures than fossil fuels, can be converted into mechanical energy with high efficiency and is clean at the point of consumpion. In addition to that electrification with renewable energy is much more efficient and therefore leads to a significant reduction in primary energy requirements, because most renewables don't have a steam cycle with high losses (fossil power plants usually have losses of 40 to 65%).

Overview

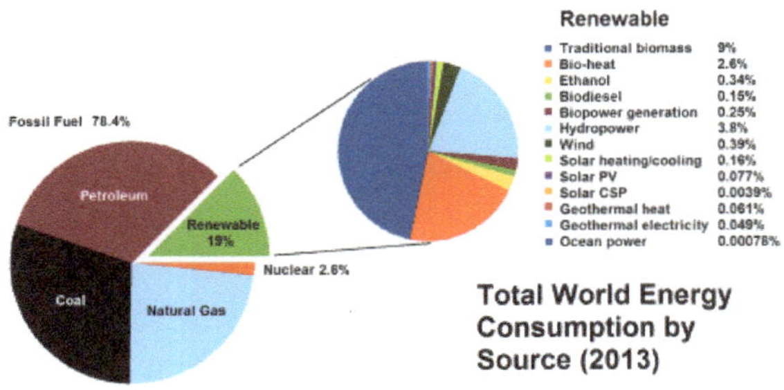

World energy consumption by source. Renewables accounted for 19% in 2012.

PlanetSolar, the world's largest solar-powered boat and the first ever solar electric vehicle to circumnavigate the globe (in 2012)

Renewable energy flows involve natural phenomena such as sunlight, wind, tides, plant growth, and geothermal heat, as the International Energy Agency explains:

Renewable energy is derived from natural processes that are replenished constantly. In its various forms, it derives directly from the sun, or from heat generated deep within the earth. Included in the definition is electricity and heat generated from solar, wind, ocean, hydropower, biomass, geothermal resources, and biofuels and hydrogen derived from renewable resources.

Renewable energy resources and significant opportunities for energy efficiency exist over wide geographical areas, in contrast to other energy sources, which are concentrated in a limited number of countries. Rapid deployment of renewable energy and energy efficiency, and technological diversification of energy sources, would result in significant energy security and economic benefits. It would also reduce environmental pollution such as air pollution caused by burning of fossil fuels and improve public health, reduce premature mortalities due to pollution and save associated health costs that amount to several hundred billion dollars annually only in the United States. Renewable energy sources, that derive their energy from the sun, either directly or indirectly, such as hydro and wind, are expected to be capable of supplying humanity energy for almost another 1 billion years, at which point the predicted increase in heat from the sun is expected to make the surface of the earth too hot for liquid water to exist.

Climate change and global warming concerns, coupled with high oil prices, peak oil, and increasing government support, are driving increasing renewable energy legislation, incentives and commercialization. New government spending, regulation and policies helped the industry weather the global financial crisis better than many other sectors. According to a 2011 projection by the International Energy Agency, solar power generators may produce most of the world's electricity within 50 years, reducing the emissions of greenhouse gases that harm the environment.

As of 2011, small solar PV systems provide electricity to a few million households, and micro-hydro configured into mini-grids serves many more. Over 44 million households use biogas made in household-scale digesters for lighting and/or cooking, and more than 166 million households rely on a new generation of more-efficient biomass cookstoves. United Nations' Secretary-General Ban Ki-moon has said that renewable energy has the ability to lift the poorest nations to new levels of prosperity. At the national level, at least 30 nations around the world already have renewable energy contributing more than 20% of energy supply. National renewable energy markets are projected to continue to grow strongly in the coming decade and beyond, and some 120 countries have various policy targets for longer-term shares of renewable energy, including a 20% target of all electricity generated for the European Union by 2020. Some countries have much higher long-term policy targets of up to 100% renewables. Outside Europe, a diverse group of 20 or more other countries target renewable energy shares in the 2020–2030 time frame that range from 10% to 50%.

Renewable energy often displaces conventional fuels in four areas: electricity generation, hot water/space heating, transportation, and rural (off-grid) energy services:

- Power generation

  Renewables account for 28.9% of the world's power generating capacity - enough to supply 23.7% of global electricity. Hydroelectric energy provides 16.6% of global electricity, while the remaining 7.1% comes from wind, geothermal, solar PV, bio-power, concentrated solar power (CSP) and ocean. By 2040, renewable energy is projected to equal coal and natural gas electricity generation. Several jurisdictions, including Denmark, Germany, the state of South Australia and some US states have achieved high integration of variable renewables. For example, in 2015 wind power met 42% of electricity demand in Denmark, 23.2% in Portugal and 15.5% in Uruguay. Interconnectors enable countries to balance electricity systems by allowing the import and export of renewable energy. Innovative hybrid systems have emerged between countries and regions.

- Heating

Solar water heating makes an important contribution to renewable heat in many countries, most notably in China, which now has 70% of the global total (180 GWth). Most of these systems are installed on multi-family apartment buildings and meet a portion of the hot water needs of an estimated 50–60 million households in China. Worldwide, total installed solar water heating systems meet a portion of the water heating needs of over 70 million households. The use of biomass for heating continues to grow as well. In Sweden, national use of biomass energy has surpassed that of oil. Direct geothermal for heating is also growing rapidly. The newest addition to Heating is from Geothermal Heat Pumps which provide both heating and cooling, and also flatten the electric demand curve and are thus an increasing national priority.

- Transportation

A bus fueled by biodiesel

Bioethanol is an alcohol made by fermentation, mostly from carbohydrates produced in sugar or starch crops such as corn, sugarcane, or sweet sorghum. Cellulosic biomass, derived from non-food sources such as trees and grasses is also being developed as a feedstock for ethanol production. Ethanol can be used as a fuel for vehicles in its pure form, but it is usually used as a gasoline additive to increase octane and improve vehicle emissions. Bioethanol is widely used in the USA and in Brazil. Biodiesel can be used as a fuel for vehicles in its pure form, but it is usually used as a diesel additive to reduce levels of particulates, carbon monoxide, and hydrocarbons from diesel-powered vehicles. Biodiesel is produced from oils or fats using transesterification and is the most common biofuel in Europe.

A solar vehicle is an electric vehicle powered completely or significantly by direct solar energy. Usually, photovoltaic (PV) cells contained in solar panels convert the sun's energy directly into electric energy. The term "solar vehicle" usually implies that solar energy is used to power all or part of a vehicle's propulsion. Solar power may be also used to provide power for communications or controls or other auxiliary functions. Solar vehicles are not sold as practical day-to-day transportation devices at present, but are primarily demonstration vehicles and engineering exercises, often sponsored by government agencies. However, indirectly solar-charged vehicles are widespread and solar boats are available commercially.

History

Prior to the development of coal in the mid 19th century, nearly all energy used was renewable. Almost without a doubt the oldest known use of renewable energy, in the form of traditional biomass to fuel fires, dates from 790,000 years ago. Use of biomass for fire did not become commonplace until many hundreds of thousands of years later, sometime between 200,000 and 400,000 years ago. Probably the second oldest usage of renewable energy is harnessing the wind in order to drive ships over water. This practice can be traced back some 7000 years, to ships on the Nile. Moving into the time of recorded history, the primary sources of traditional renewable energy were human labor, animal power, water power, wind, in grain crushing windmills, and firewood, a traditional biomass. A graph of energy use in the United States up until 1900 shows oil and natural gas with about the same importance in 1900 as wind and solar played in 2010.

In the 1860s and '70s there were already fears that civilization would run out of fossil fuels and the need was felt for a better source. In 1873 Professor Augustine Mouchot wrote:

The time will arrive when the industry of Europe will cease to find those natural resources, so necessary for it. Petroleum springs and coal mines are not inexhaustible but are rapidly diminishing in many places. Will man, then, return to the power of water and wind? Or will he emigrate where the most powerful source of heat sends its rays to all? History will show what will come.

In 1885, Werner von Siemens, commenting on the discovery of the photovoltaic effect in the solid state, wrote:

In conclusion, I would say that however great the scientific importance of this discovery may be, its practical value will be no less obvious when we reflect that the supply of solar energy is both without limit and without cost, and that it will continue to pour down upon us for countless ages after all the coal deposits of the earth have been exhausted and forgotten.

Max Weber mentioned the end of fossil fuel in the concluding paragraphs of his Die protestantische Ethik und der Geist des Kapitalismus, published in 1905.

Development of solar engines continued until the outbreak of World War I. The importance of solar energy was recognized in a 1911 *Scientific American* article: "in the far distant future, natural fuels having been exhausted [solar power] will remain as the only means of existence of the human race".

The theory of peak oil was published in 1956. In the 1970s environmentalists promoted the development of renewable energy both as a replacement for the eventual depletion of oil, as well as for an escape from dependence on oil, and the first electricity generating wind turbines appeared. Solar had long been used for heating and cooling, but solar panels were too costly to build solar farms until 1980.

The IEA 2014 World Energy Outlook projects a growth of renewable energy supply from 1,700 gigawatts in 2014 to 4,550 gigawatts in 2040. Fossil fuels received about $550 billion in subsidies in 2013, compared to $120 billion for all renewable energies.

Mainstream Technologies

Wind Power

The 845 MW Shepherds Flat Wind Farm near Arlington, Oregon, USA

Airflows can be used to run wind turbines. Modern utility-scale wind turbines range from around 600 kW to 5 MW of rated power, although turbines with rated output of 1.5–3 MW have become the most common for commercial use; the power available from the wind is a function of the cube of the wind speed, so as wind speed increases, power output increases up to the maximum output for the particular turbine. Areas where winds are stronger and more constant, such as offshore and high altitude sites, are preferred locations for wind farms. Typically full load hours of wind turbines vary between 16 and 57 percent annually, but might be higher in particularly favorable offshore sites.

Globally, the long-term technical potential of wind energy is believed to be five times total current global energy production, or 40 times current electricity demand, assuming all practical barriers needed were overcome. This would require wind turbines to be installed over large areas, particularly in areas of higher wind resources, such as offshore. As offshore wind speeds average ~90% greater than that of land, so offshore resources can contribute substantially more energy than land stationed turbines. In 2014 global wind generation was 706 terawatt-hours or 3% of the worlds total electricity.

Hydropower

The Three Gorges Dam on the Yangtze River in China

In 2015 hydropower generated 16.6% of the worlds total electricity and 70% of all renewable electricity. Since water is about 800 times denser than air, even a slow flowing stream of water, or moderate sea swell, can yield considerable amounts of energy. There are many forms of water energy:

- Historically hydroelectric power came from constructing large hydroelectric dams and reservoirs, which are still popular in third world countries. The largest of which is the Three Gorges Dam(2003) in China and the Itaipu Dam(1984) built by Brazil and Paraguay.
- Small hydro systems are hydroelectric power installations that typically produce up to 50 MW of power. They are often used on small rivers or as a low impact development on larger rivers. China is the largest producer of hydroelectricity in the world and has more than 45,000 small hydro installations.
- Run-of-the-river hydroelectricity plants derive kinetic energy from rivers without the creation of a large reservoir. This style of generation may still produce a large amount of electricity, such as the Chief Joseph Dam on the Columbia river in the United States.

Hydropower is produced in 150 countries, with the Asia-Pacific region generating 32 percent of global hydropower in 2010. For counties having the largest percentage of electricity from renewables, the top 50 are primarily hydroelectric. China is the largest hydroelectricity producer, with 721 terawatt-hours of production in 2010, representing around 17 percent of domestic electricity use. There are now three hydroelectricity stations larger than 10 GW: the Three Gorges Dam in China, Itaipu Dam across the Brazil/Paraguay border, and Guri Dam in Venezuela.

Wave power, which captures the energy of ocean surface waves, and tidal power, converting the energy of tides, are two forms of hydropower with future potential; however, they are not yet widely employed commercially. A demonstration project operated by the Ocean Renewable Power Company on the coast of Maine, and connected to the grid, harnesses tidal power from the Bay of Fundy, location of world's highest tidal flow. Ocean thermal energy conversion, which uses the temperature difference between cooler deep and warmer surface waters, has currently no economic feasibility.

Solar Energy

Satellite image of the 550-megawatt Topaz Solar Farm in California, USA

Solar energy, radiant light and heat from the sun, is harnessed using a range of ever-evolving technologies such as solar heating, photovoltaics, concentrated solar power (CSP), concentrator photovoltaics (CPV), solar architecture and artificial photosynthesis. Solar technologies are broadly characterized as either passive solar or active solar depending on the way they capture, convert and distribute solar energy. Passive solar techniques include orienting a building to the Sun, selecting materials with favorable thermal mass or light dispersing properties, and designing spaces that naturally circulate air. Active solar technologies encompass solar thermal energy, using solar collectors for heating, and solar power, converting sunlight into electricity either directly using photovoltaics (PV), or indirectly using concentrated solar power (CSP).

A photovoltaic system converts light into electrical direct current (DC) by taking advantage of the photoelectric effect. Solar PV has turned into a multi-billion, fast-growing industry, continues to improve its cost-effectiveness, and has the most potential of any renewable technologies together with CSP. Concentrated solar power (CSP) systems use lenses or mirrors and tracking systems to focus a large area of sunlight into a small beam. Commercial concentrated solar power plants were first developed in the 1980s. CSP-Stirling has by far the highest efficiency among all solar energy technologies.

In 2011, the International Energy Agency said that "the development of affordable, inexhaustible and clean solar energy technologies will have huge longer-term benefits. It will increase countries' energy security through reliance on an indigenous, inexhaustible and mostly import-independent resource, enhance sustainability, reduce pollution, lower the costs of mitigating climate change, and keep fossil fuel prices lower than otherwise. These advantages are global. Hence the additional costs of the incentives for early deployment should be considered learning investments; they must be wisely spent and need to be widely shared". In 2014 global solar generation was 186 terawatt-hours, slightly less than 1% of the worlds total grid electricity.

Geothermal Energy

Steam rising from the Nesjavellir Geothermal Power Station in Iceland

High Temperature Geothermal energy is from thermal energy generated and stored in the Earth. Thermal energy is the energy that determines the temperature of matter. Earth's geothermal energy originates from the original formation of the planet and from radioactive decay of minerals (in currently uncertain but possibly roughly equal proportions). The geothermal gradient, which is the difference in temperature between the core of the planet and its surface, drives a continuous conduction of thermal energy in the form of heat from the core to the surface. The adjective *geothermal* originates from the Greek roots *geo*, meaning earth, and *thermos*, meaning heat.

The heat that is used for geothermal energy can be from deep within the Earth, all the way down to Earth's core – 4,000 miles (6,400 km) down. At the core, temperatures may reach over 9,000 °F (5,000 °C). Heat conducts from the core to surrounding rock. Extremely high temperature and pressure cause some rock to melt, which is commonly known as magma. Magma convects upward since it is lighter than the solid rock. This magma then heats rock and water in the crust, sometimes up to 700 °F (371 °C).

From hot springs, geothermal energy has been used for bathing since Paleolithic times and for space heating since ancient Roman times, but it is now better known for electricity generation.

Low Temperature Geothermal refers to the use of the outer crust of the earth as a Thermal Battery to facilitate Renewable thermal energy for heating and cooling buildings, and other refrigeration and industrial uses. In this form of Geothermal, a Geothermal Heat Pump and Ground-coupled heat exchanger are used together to move heat energy into the earth (for cooling) and out of the earth (for heating) on a varying seasonal basis. Low temperature Geothermal (generally referred to as "GHP") is an increasingly important renewable technology because it both reduces total annual energy loads associated with heating and cooling, and it also flattens the electric demand curve eliminating the extreme summer and winter peak electric supply requirements. Thus Low Temperature Geothermal/GHP is becoming an increasing national priority with multiple tax credit support and focus as part of the ongoing movement toward Net Zero Energy. New York City has even just passed a law to require GHP anytime is shown to be economical with 20 year financing including the Socialized Cost of Carbon.

Bio Energy

Biomass is biological material derived from living, or recently living organisms. It most often refers to plants or plant-derived materials which are specifically called lignocellulosic biomass. As an energy source, biomass can either be used directly via combustion to produce heat, or indirectly after converting it to various forms of biofuel. Conversion of biomass to biofuel can be achieved by different methods which are broadly classified into: *thermal*, *chemical*, and *biochemical* methods. Wood remains the largest biomass energy source today; examples include forest residues – such as dead trees, branches and tree stumps –, yard clippings, wood chips and even municipal solid waste. In the second sense, biomass includes plant or animal matter that can be converted into fibers or other industrial chemicals, including biofuels. Industrial biomass can be grown from numerous types of plants, including miscanthus, switchgrass, hemp, corn, poplar, willow, sorghum, sugarcane, bamboo, and a variety of tree species, ranging from eucalyptus to oil palm (palm oil).

Plant energy is produced by crops specifically grown for use as fuel that offer high biomass output per hectare with low input energy. Some examples of these plants are wheat, which typically yield 7.5–8 tonnes of grain per hectare, and straw, which typically yield 3.5–5 tonnes per hectare in the UK. The grain can be used for liquid transportation fuels while the straw can be burned to produce heat or electricity. Plant biomass can also be degraded from cellulose to glucose through a series of chemical treatments, and the resulting sugar can then be used as a first generation biofuel.

Biomass can be converted to other usable forms of energy like methane gas or transportation fuels like ethanol and biodiesel. Rotting garbage, and agricultural and human waste, all release methane gas – also called landfill gas or biogas. Crops, such as corn and sugarcane, can be fermented to produce the transportation fuel, ethanol. Biodiesel, another transportation fuel, can be produced from left-over food products like vegetable oils and animal fats. Also, biomass to liquids (BTLs) and cellulosic ethanol are still under research. There is a great deal of research involving algal fuel or algae-derived biomass due to the fact that it's a non-food resource and can be produced at rates 5 to 10 times those of other types of land-based agriculture, such as corn and soy. Once harvested, it can be fermented to produce biofuels such as ethanol, butanol, and methane, as well as biodiesel and hydrogen. The biomass used for electricity generation varies by region. Forest by-products, such as wood residues, are common in the United States. Agricultural waste is common in Mauritius (sugar cane residue) and Southeast Asia (rice husks). Animal husbandry residues, such as poultry litter, are common in the United Kingdom.

Biofuels include a wide range of fuels which are derived from biomass. The term covers solid, liquid, and gaseous fuels. Liquid biofuels include bioalcohols, such as bioethanol, and oils, such as biodiesel. Gaseous biofuels include biogas, landfill gas and synthetic gas. Bioethanol is an alcohol made by fermenting the sugar components of plant materials and it is made mostly from sugar and starch crops. These include maize, sugarcane and, more recently, sweet sorghum. The latter crop is particularly suitable for growing in dryland conditions, and is being investigated by International Crops Research Institute for the Semi-Arid Tropics for its potential to provide fuel, along with food and animal feed, in arid parts of Asia and Africa.

With advanced technology being developed, cellulosic biomass, such as trees and grasses, are also used as feedstocks for ethanol production. Ethanol can be used as a fuel for vehicles in its pure form, but it is usually used as a gasoline additive to increase octane and improve vehicle emissions. Bioethanol is widely used in the United States and in Brazil. The energy costs for producing bio-ethanol are almost equal to, the energy yields from bio-ethanol. However, according to the European Environment Agency, biofuels do not address global warming concerns. Biodiesel is made from vegetable oils, animal fats or recycled greases. It can be used as a fuel for vehicles in its pure form, or more commonly as a diesel additive to reduce levels of particulates, carbon monoxide, and hydrocarbons from diesel-powered vehicles. Biodiesel is produced from oils or fats using transesterification and is the most common biofuel in Europe. Biofuels provided 2.7% of the world's transport fuel in 2010.

Biomass, biogas and biofuels are burned to produce heat/power and in doing so harm the environment. Pollutants such as sulphurous oxides (SO_x), nitrous oxides (NO_x), and particulate matter (PM) are produced from the combustion of biomass; the World Health Organisation estimates that 7 million premature deaths are caused each year by air pollution. Biomass combustion is a major contributor. The life cycle of the plants is sustainable, the lives of people less so.

Energy Storage

Energy storage is a collection of methods used to store electrical energy on an electrical power grid, or off it. Electrical energy is stored during times when production (especially from intermittent power plants such as renewable electricity sources such as wind power, tidal power, solar power) exceeds consumption, and returned to the grid when production falls below consumption. Water pumped into a hydroelectric dam is the largest form of power storage.

Market and Industry Trends

Growth of Renewables

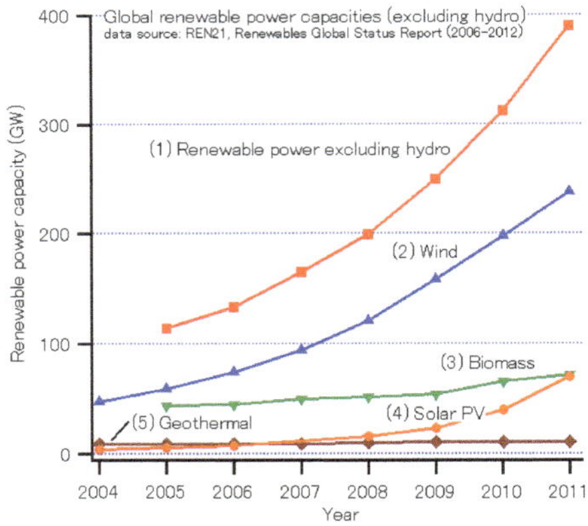

Global growth of renewables through to 2011

From the end of 2004, worldwide renewable energy capacity grew at rates of 10–60% annually for many technologies. For wind power and many other renewable technologies, growth accelerated in 2009 relative to the previous four years. More wind power capacity was added during 2009 than any other renewable technology. However, grid-connected PV increased the fastest of all renewables technologies, with a 60% annual average growth rate. In 2010, renewable power constituted about a third of the newly built power generation capacities.

Projections vary, but scientists have advanced a plan to power 100% of the world's energy with wind, hydroelectric, and solar power by the year 2030.

According to a 2011 projection by the International Energy Agency, solar power generators may produce most of the world's electricity within 50 years, reducing the emissions of greenhouse gases that harm the environment. Cedric Philibert, senior analyst in the renewable energy division at the IEA said: "Photovoltaic and solar-thermal plants may meet most of the world's demand for electricity by 2060 – and half of all energy needs – with wind, hydropower and biomass plants supplying much of the remaining generation". "Photovoltaic and concentrated solar power together can become the major source of electricity", Philibert said.

In 2014 global wind power capacity expanded 16% to 369,553 MW. Yearly wind energy production is also growing rapidly and has reached around 4% of worldwide electricity usage, 11.4% in the EU, and it is widely used in Asia, and the United States. In 2015, worldwide installed photovoltaics capacity increased to 227 gigawatts (GW), sufficient to supply 1 percent of global electricity demands. Solar thermal energy stations operate in the USA and Spain, and as of 2016, the largest of these is the 392 MW Ivanpah Solar Electric Generating System in California. The world's largest geothermal power installation is The Geysers in California, with a rated capacity of 750 MW. Brazil has one of the largest renewable energy programs in the world, involving production of ethanol fuel from sugar cane, and ethanol now provides 18% of the country's automotive fuel. Ethanol fuel is also widely available in the USA.

Selected renewable energy global indicators	2008	2009	2010	2011	2012	2013	2014	2015
Investment in new renewable capacity (annual) (10^9 USD)	182	178	237	279	256	232	270	285
Renewables power capacity (existing) (GWe)	1,140	1,230	1,320	1,360	1,470	1,578	1,712	1,849
Hydropower capacity (existing) (GWe)	885	915	945	970	990	1,018	1,055	1,064
Wind power capacity (existing) (GWe)	121	159	198	238	283	319	370	433
Solar PV capacity (grid-connected) (GWe)	16	23	40	70	100	138	177	227
Solar hot water capacity (existing) (GWth)	130	160	185	232	255	373	406	435
Ethanol production (annual) (10^9 litres)	67	76	86	86	83	87	94	98
Biodiesel production (annual) (10^9 litres)	12	17.8	18.5	21.4	22.5	26	29.7	30
Countries with policy targets for renewable energy use	79	89	98	118	138	144	164	173
Source: The Renewable Energy Policy Network for the 21st Century (REN21)–Global Status Report								

Economic Trends

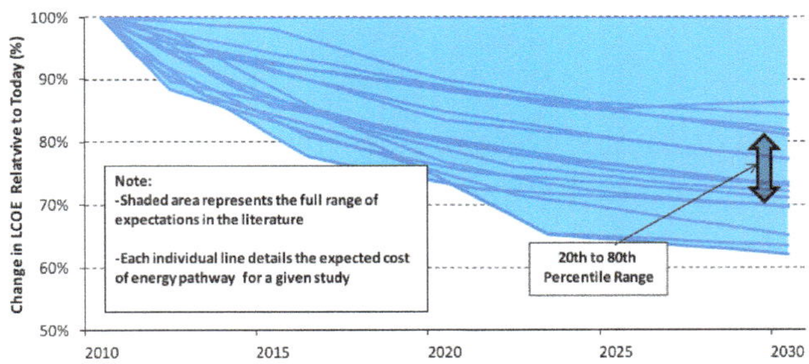

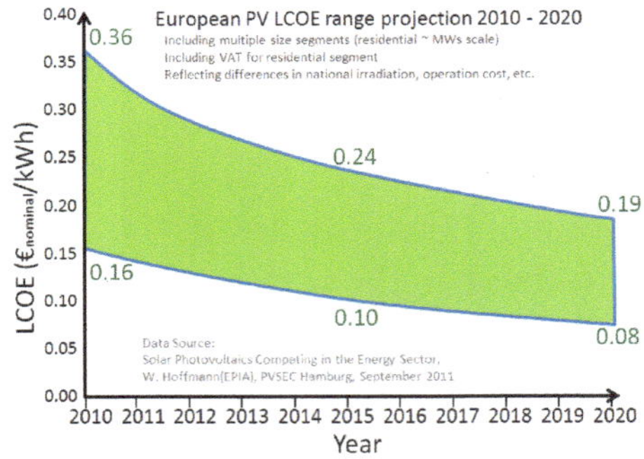

Projection of levelized cost for wind in the U.S. (left) and solar power in Europe

Renewable energy technologies are getting cheaper, through technological change and through the benefits of mass production and market competition. A 2011 IEA report said: "A portfolio of renewable energy technologies is becoming cost-competitive in an increasingly broad range of circumstances, in some cases providing investment opportunities without the need for specific economic support," and added that "cost reductions in critical technologies, such as wind and solar, are set to continue."

Hydro-electricity and geothermal electricity produced at favourable sites are now the cheapest way to generate electricity. Renewable energy costs continue to drop, and the levelised cost of electricity (LCOE) is declining for wind power, solar photovoltaic (PV), concentrated solar power (CSP) and some biomass technologies. Renewable energy is also the most economic solution for new grid-connected capacity in areas with good resources. As the cost of renewable power falls, the scope of economically viable applications increases. Renewable technologies are now often the most economic solution for new generating capacity. Where "oil-fired generation is the predominant power generation source (e.g. on islands, off-grid and in some countries) a lower-cost renewable solution almost always exists today". A series of studies by the US National Renewable Energy Laboratory modeled the "grid in the Western US under a number of different scenarios where intermittent renewables accounted for 33 percent of the total power." In the models, inefficiencies in cycling the fossil fuel plants to compensate for the variation in solar and wind energy resulted in an additional cost of "between $0.47 and $1.28 to each MegaWatt hour generated"; however, the savings in the cost of the fuels saved "adds up to $7 billion, meaning the added costs are, at most, two percent of the savings."

Hydroelectricity

Only 25% of the worlds estimated hydroelectric potential of 14,000 TWh/year has been developed, with Africa, Asia and Latin America having the greatest potential. The Three Gorges Dam in Hubei, China, has the world's largest instantaneous generating capacity (22,500 MW), with the Itaipu Dam in Brazil/Paraguay in second place (14,000 MW). The Three Gorges Dam is operated jointly with the much smaller Gezhouba Dam (3,115 MW). As of 2012, the total generating capacity of this two-dam complex is 25,615 MW.

In 2008, this complex generated 98 TWh of electricity (81 TWh from the Three Gorges Dam and 17 TWh from the Gezhouba Dam), which is 3% more power in one year than the 95 TWh generated by Itaipu in 2008.

Wind Power Development

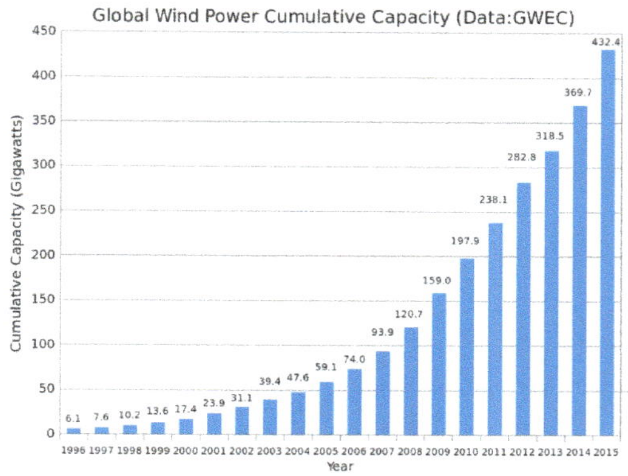

Worldwide growth of wind capacity (1996–2014)

Four offshore wind farms are in the Thames Estuary area: Kentish Flats, Gunfleet Sands, Thanet and London Array. The latter is the largest in the world as of April 2013.

Wind power is widely used in Europe, China, and the United States. From 2004 to 2014, worldwide installed capacity of wind power has been growing from 47 GW to 369 GW—a more than sevenfold increase within 10 years with 2014 breaking a new record in global installations (51 GW). As of the end of 2014, China, the United States and Germany combined accounted for half of total global capacity. Several other countries have achieved relatively high levels of wind power penetration, such as 21% of stationary electricity production in Denmark, 18% in Portugal, 16% in Spain, and 14% in Ireland in 2010 and have since continued to expand their installed capacity. More than 80 countries around the world are using wind power on a commercial basis.

- Offshore wind power

As of 2014, offshore wind power amounted to 8,771 megawatt of global installed capacity. Although offshore capacity doubled within three years (from 4,117 MW in 2011), it accounted for only 2.3% of the total wind power capacity. The United Kingdom is the undisputed leader of offshore power with half of the world's installed capacity ahead of Denmark, Germany, Belgium and China.

- List of offshore and onshore wind farms

  As of 2012, the Alta Wind Energy Center (California, 1,020 MW) is the world's largest wind farm. The London Array (630 MW) is the largest offshore wind farm in the world. The United Kingdom is the world's leading generator of offshore wind power, followed by Denmark. There are several large offshore wind farms operational and under construction and these include Anholt (400 MW), BARD (400 MW), Clyde (548 MW), Fântânele-Cogealac (600 MW), Greater Gabbard (500 MW), Lincs (270 MW), London Array (630 MW), Lower Snake River (343 MW), Macarthur (420 MW), Shepherds Flat (845 MW), and the Sheringham Shoal (317 MW).

Solar Thermal

The 377 MW Ivanpah Solar Electric Generating System with all three towers under load, Feb 2014. Taken from I-15.

Solar Towers of the PS10 and PS20 solar thermal plants in Spain

The United States conducted much early research in photovoltaics and concentrated solar power. The U.S. is among the top countries in the world in electricity generated by the Sun and several of the world's largest utility-scale installations are located in the desert Southwest.

The oldest solar thermal power plant in the world is the 354 megawatt (MW) SEGS thermal power plant, in California. The Ivanpah Solar Electric Generating System is a solar thermal power project in the California Mojave Desert, 40 miles (64 km) southwest of Las Vegas, with a gross capacity of 377 MW. The 280 MW Solana Generating Station is a solar power plant near Gila Bend, Arizona, about 70 miles (110 km) southwest of Phoenix, completed in 2013. When commissioned it was the largest parabolic trough plant in the world and the first U.S. solar plant with molten salt thermal energy storage.

The solar thermal power industry is growing rapidly with 1.3 GW under construction in 2012 and more planned. Spain is the epicenter of solar thermal power development with 873 MW under construction, and a further 271 MW under development. In the United States, 5,600 MW of solar thermal power projects have been announced. Several power plants have been constructed in the Mojave Desert, Southwestern United States. The Ivanpah Solar Power Facility being the most recent. In developing countries, three World Bank projects for integrated solar thermal/combined-cycle gas-turbine power plants in Egypt, Mexico, and Morocco have been approved.

Photovoltaic Development

Photovoltaics (PV) uses solar cells assembled into solar panels to convert sunlight into electricity. It's a fast-growing technology doubling its worldwide installed capacity every couple of years. PV systems range from small, residential and commercial rooftop or building integrated installations, to large utility-scale photovoltaic power station. The predominant PV technology is crystalline silicon, while thin-film solar cell technology accounts for about 10 percent of global photovoltaic deployment. In recent years, PV technology has improved its electricity generating efficiency, reduced the installation cost per watt as well as its energy payback time, and has reached grid parity in at least 30 different markets by 2014. Financial institutions are predicting a second solar "gold rush" in the near future.

At the end of 2014, worldwide PV capacity reached at least 177,000 megawatts. Photovoltaics grew fastest in China, followed by Japan and the United States, while Germany remains the world's largest overall producer of photovoltaic power, contributing about 7.0 percent to the overall electricity generation. Italy meets 7.9 percent of its electricity demands with photovoltaic power—the highest share worldwide. For 2015, global cumulative capacity is forecasted to increase by more than 50 gigawatts (GW). By 2018, worldwide capacity is projected to reach as much as 430 gigawatts. This corresponds to a tripling within five years. Solar power is forecasted to become the world's largest source of electricity by 2050, with solar photovoltaics and concentrated solar power contributing 16% and 11%, respectively. This requires an increase of installed PV capacity to 4,600 GW, of which more than half is expected to be deployed in China and India.

Photovoltaic Power Stations

Solar panels at the 550 MW Topaz Solar Farm

Nellis Solar Power Plant, photovoltaic power plant in Nevada, USA

Commercial concentrated solar power plants were first developed in the 1980s. As the cost of solar electricity has fallen, the number of grid-connected solar PV systems has grown into the millions and utility-scale solar power stations with hundreds of megawatts are being built. Solar PV is rapidly becoming an inexpensive, low-carbon technology to harness renewable energy from the Sun.

Many solar photovoltaic power stations have been built, mainly in Europe, China and the USA. The 579 MW Solar Star, in the United States, is the world's largest PV power station.

Many of these plants are integrated with agriculture and some use tracking systems that follow the sun's daily path across the sky to generate more electricity than fixed-mounted systems. There are no fuel costs or emissions during operation of the power stations.

However, when it comes to renewable energy systems and PV, it is not just large systems that matter. Building-integrated photovoltaics or "onsite" PV systems use existing land and structures and generate power close to where it is consumed.

Biofuel Development

Brazil produces bioethanol made from sugarcane available throughout the country. A typical gas station with dual fuel service is marked "A" for alcohol (ethanol) and "G" for gasoline.

Biofuels provided 3% of the world's transport fuel in 2010. Mandates for blending biofuels exist in 31 countries at the national level and in 29 states/provinces. According to the International Energy Agency, biofuels have the potential to meet more than a quarter of world demand for transportation fuels by 2050.

Since the 1970s, Brazil has had an ethanol fuel program which has allowed the country to become the world's second largest producer of ethanol (after the United States) and the world's largest exporter. Brazil's ethanol fuel program uses modern equipment and cheap sugarcane as feedstock, and the residual cane-waste (bagasse) is used to produce heat and power. There are no longer light vehicles in Brazil running on pure gasoline. By the end of 2008 there were 35,000 filling stations throughout Brazil with at least one ethanol pump. Unfortunately, Operation Car Wash has seriously eroded public trust in oil companies and has implicated several high ranking Brazilian officials.

Nearly all the gasoline sold in the United States today is mixed with 10% ethanol, and motor vehicle manufacturers already produce vehicles designed to run on much higher ethanol blends. Ford, Daimler AG, and GM are among the automobile companies that sell "flexible-fuel" cars, trucks, and minivans that can use gasoline and ethanol blends ranging from pure gasoline up to 85% ethanol. By mid-2006, there were approximately 6 million ethanol compatible vehicles on U.S. roads.

Geothermal Development

Geothermal power is cost effective, reliable, sustainable, and environmentally friendly, but has historically been limited to areas near tectonic plate boundaries. Recent technological advances have expanded the range and size of viable resources, especially for applications such as home heating, opening a potential for widespread exploitation. Geothermal wells release greenhouse gases trapped deep within the earth, but these emissions are much lower per energy unit than those of fossil fuels. As a result, geothermal power has the potential to help mitigate global warming if widely deployed in place of fossil fuels.

The International Geothermal Association (IGA) has reported that 10,715 MW of geothermal power in 24 countries is online, which is expected to generate 67,246 GWh of electricity in 2010. This represents a 20% increase in geothermal power online capacity since 2005. IGA projects this will grow to 18,500 MW by 2015, due to the large number of projects presently under consideration, often in areas previously assumed to have little exploitable resource.

In 2010, the United States led the world in geothermal electricity production with 3,086 MW of installed capacity from 77 power plants; the largest group of geothermal power plants in the world is located at The Geysers, a geothermal field in California. The Philippines follows the US as the second highest producer of geothermal power in the world, with 1,904 MW of capacity online; geothermal power makes up approximately 18% of the country's electricity generation.

Developing Countries

Solar cookers use sunlight as energy source for outdoor cooking

Renewable energy technology has sometimes been seen as a costly luxury item by critics, and affordable only in the affluent developed world. This erroneous view has persisted for many years, but 2015 was the first year when investment in non-hydro renewables, was higher in developing countries, with $156 billion invested, mainly in China, India, and Brazil.

Renewable energy can be particularly suitable for developing countries. In rural and remote areas, transmission and distribution of energy generated from fossil fuels can be difficult and expensive. Producing renewable energy locally can offer a viable alternative.

Technology advances are opening up a huge new market for solar power: the approximately 1.3 billion people around the world who don't have access to grid electricity. Even though they are typically very poor, these people have to pay far more for lighting than people in rich countries because they use inefficient kerosene lamps. Solar power costs half as much as lighting with kerosene. As of 2010, an estimated 3 million households get power from small solar PV systems. Kenya is the world leader in the number of solar power systems installed per capita. More than 30,000 very small solar panels, each producing 12 to 30 watts, are sold in Kenya annually. Some Small Island Developing States (SIDS) are also turning to solar power to reduce their costs and increase their sustainability.

Micro-hydro configured into mini-grids also provide power. Over 44 million households use biogas made in household-scale digesters for lighting and/or cooking, and more than 166 million households rely on a new generation of more-efficient biomass cookstoves. Clean liquid fuel sourced from renewable feedstocks are used for cooking and lighting in energy-poor areas of the developing world. Alcohol fuels (ethanol and methanol) can be produced sustainably from non-food sugary, starchy, and cellulostic feedstocks. Project Gaia, Inc. and CleanStar Mozambique are implementing clean cooking programs with liquid ethanol stoves in Ethiopia, Kenya, Nigeria and Mozambique.

Renewable energy projects in many developing countries have demonstrated that renewable energy can directly contribute to poverty reduction by providing the energy needed for creating businesses and employment. Renewable energy technologies can also make indirect contributions to alleviating poverty by providing energy for cooking, space heating, and lighting. Renewable energy can also contribute to education, by providing electricity to schools.

Industry and Policy Trends

U.S. President Barack Obama's American Recovery and Reinvestment Act of 2009 includes more than $70 billion in direct spending and tax credits for clean energy and associated transportation programs. Leading renewable energy companies include First Solar, Gamesa, GE Energy, Hanwha Q Cells, Sharp Solar, Siemens, SunOpta, Suntech Power, and Vestas.

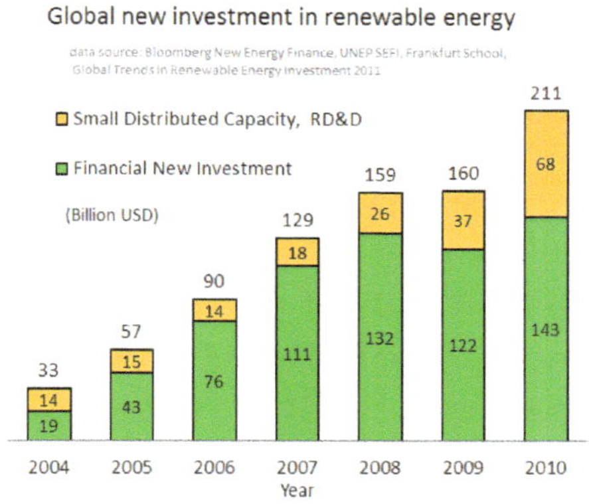

Global New Investments in Renewable Energy

Many national, state, and local governments have also created green banks. A green bank is a quasi-public financial institution that uses public capital to leverage private investment in clean energy technologies. Green banks use a variety of financial tools to bridge market gaps that hinder the deployment of clean energy.

The military has also focused on the use of renewable fuels for military vehicles. Unlike fossil fuels, renewable fuels can be produced in any country, creating a strategic advantage. The US military has already committed itself to have 50% of its energy consumption come from alternative sources.

The International Renewable Energy Agency (IRENA) is an intergovernmental organization for promoting the adoption of renewable energy worldwide. It aims to provide concrete policy advice and facilitate capacity building and technology transfer. IRENA was formed on 26 January 2009, by 75 countries signing the charter of IRENA. As of March 2010, IRENA has 143 member states who all are considered as founding members, of which 14 have also ratified the statute.

As of 2011, 119 countries have some form of national renewable energy policy target or renewable support policy. National targets now exist in at least 98 countries. There is also a wide range of policies at state/provincial and local levels.

United Nations' Secretary-General Ban Ki-moon has said that renewable energy has the ability to lift the poorest nations to new levels of prosperity. In October 2011, he "announced the creation of a high-level group to drum up support for energy access, energy efficiency and greater use of renewable energy. The group is to be co-chaired by Kandeh Yumkella, the chair of UN Energy and director general of the UN Industrial Development Organisation, and Charles Holliday, chairman of Bank of America".

100% Renewable Energy

The incentive to use 100% renewable energy, for electricity, transport, or even total primary energy supply globally, has been motivated by global warming and other ecological as well as economic concerns. The Intergovernmental Panel on Climate Change has said that there are few fundamental technological limits to integrating a portfolio of renewable energy technologies to meet most of total global energy demand. Renewable energy use has grown much faster than even advocates anticipated. At the national level, at least 30 nations around the world already have renewable energy contributing more than 20% of energy supply. Also, Professors S. Pacala and Robert H. Socolow have developed a series of "stabilization wedges" that can allow us to maintain our quality of life while avoiding catastrophic climate change, and "renewable energy sources," in aggregate, constitute the largest number of their "wedges".

Using 100% renewable energy was first suggested in a Science paper published in 1975 by Danish physicist Bent Sørensen. It was followed by several other proposals, until in 1998 the first detailed analysis of scenarios with very high shares of renewables were published. These were followed by the first detailed 100% scenarios. In 2006 a PhD thesis was published by Czisch in which it was shown that in a 100% renewable scenario energy supply could match demand in every hour of the year in Europa and North

Africa. In the same year Danish Energy professor Henrik Lund published a first paper in which he addresses the optimal combination of renewables, which was followed by several other papers on the transition to 100% renewable energy in Denmark. Since then Lund has been publishing several papers on 100% renewable energy. After 2009 publications began to rise steeply, covering 100% scenarios for countries in Europa, America, Australia and other parts of the world.

In 2011 Mark Z. Jacobson, professor of civil and environmental engineering at Stanford University, and Mark Delucchi published a study on 100% renewable global energy supply in the journal Energy Policy. They found producing all new energy with wind power, solar power, and hydropower by 2030 is feasible and existing energy supply arrangements could be replaced by 2050. Barriers to implementing the renewable energy plan are seen to be "primarily social and political, not technological or economic". They also found that energy costs with a wind, solar, water system should be similar to today's energy costs.

Similarly, in the United States, the independent National Research Council has noted that "sufficient domestic renewable resources exist to allow renewable electricity to play a significant role in future electricity generation and thus help confront issues related to climate change, energy security, and the escalation of energy costs ... Renewable energy is an attractive option because renewable resources available in the United States, taken collectively, can supply significantly greater amounts of electricity than the total current or projected domestic demand." .

The most significant barriers to the widespread implementation of large-scale renewable energy and low carbon energy strategies are primarily political and not technological. According to the 2013 *Post Carbon Pathways* report, which reviewed many international studies, the key roadblocks are: climate change denial, the fossil fuels lobby, political inaction, unsustainable energy consumption, outdated energy infrastructure, and financial constraints.

Emerging Technologies

Other renewable energy technologies are still under development, and include cellulosic ethanol, hot-dry-rock geothermal power, and marine energy. These technologies are not yet widely demonstrated or have limited commercialization. Many are on the horizon and may have potential comparable to other renewable energy technologies, but still depend on attracting sufficient attention and research, development and demonstration (RD&D) funding.

There are numerous organizations within the academic, federal, and commercial sectors conducting large scale advanced research in the field of renewable energy. This research spans several areas of focus across the renewable energy spectrum. Most of the research is targeted at improving efficiency and increasing overall energy yields. Multiple federally supported research organizations have focused on renewable energy in recent years. Two of the most prominent of these labs are Sandia National Laboratories and the National Renewable Energy Laboratory (NREL), both of which are funded by the United States Department of Energy and supported by various corporate partners. Sandia has a total budget of $2.4 billion while NREL has a budget of $375 million.

- Enhanced geothermal system

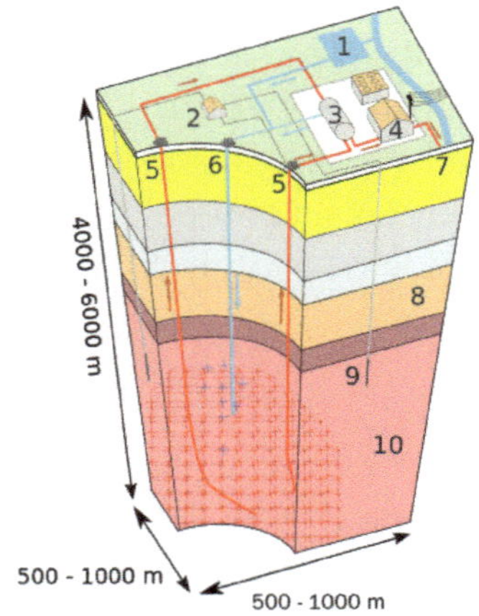

Enhanced geothermal system

Enhanced geothermal systems (EGS) are a new type of geothermal power technologies that do not require natural convective hydrothermal resources. The vast majority of geothermal energy within drilling reach is in dry and non-porous rock. EGS technologies "enhance" and/or create geothermal resources in this "hot dry rock (HDR)" through hydraulic stimulation. EGS and HDR technologies, like hydrothermal geothermal, are expected to be baseload resources which produce power 24 hours a day like a fossil plant. Distinct from hydrothermal, HDR and EGS may be feasible anywhere in the world, depending on the economic limits of drill depth. Good locations are over deep granite covered by a thick (3–5 km) layer of insulating sediments which slow heat loss. There are HDR and EGS systems currently being developed and tested in France, Australia, Japan, Germany, the U.S. and Switzerland. The largest EGS project in the world is a 25 megawatt demonstration plant currently being developed in the Cooper Basin, Australia. The Cooper Basin has the potential to generate 5,000–10,000 MW.

- Cellulosic ethanol

Several refineries that can process biomass and turn it into ethanol are built by companies such as Iogen, POET, and Abengoa, while other companies such as the Verenium Corporation, Novozymes, and Dyadic International are producing enzymes which could enable future commercialization. The shift from food crop feedstocks to waste residues and native grasses offers significant opportunities for a range of players, from farmers to biotechnology firms, and from project developers to investors.

- Marine energy

Rance Tidal Power Station, France

Marine energy (also sometimes referred to as ocean energy) refers to the energy carried by ocean waves, tides, salinity, and ocean temperature differences. The movement of water in the world's oceans creates a vast store of kinetic energy, or energy in motion. This energy can be harnessed to generate electricity to power homes, transport and industries. The term marine energy encompasses both wave power – power from surface waves, and tidal power – obtained from the kinetic energy of large bodies of moving water. Reverse electrodialysis (RED) is a technology for generating electricity by mixing fresh river water and salty sea water in large power cells designed for this purpose; as of 2016 it is being tested at a small scale (50 kW). Offshore wind power is not a form of marine energy, as wind power is derived from the wind, even if the wind turbines are placed over water. The oceans have a tremendous amount of energy and are close to many if not most concentrated populations. Ocean energy has the potential of providing a substantial amount of new renewable energy around the world.

#	Station	Country	Location	Capacity	Refs
1.	Sihwa Lake Tidal Power Station	South Korea	37°18′47″N 126°36′46″E37.31306°N 126.61278°E	254 MW	
2.	Rance Tidal Power Station	France	48°37′05″N 02°01′24″W48.61806°N 2.02333°W	240 MW	
3.	Annapolis Royal Generating Station	Canada	44°45′07″N 65°30′40″W44.75194°N 65.51111°W	20 MW	

- Experimental solar power

  Concentrated photovoltaics (CPV) systems employ sunlight concentrated onto photovoltaic surfaces for the purpose of electricity generation. Thermoelectric, or "thermovoltaic" devices convert a temperature difference between dissimilar materials into an electric current.

- Floating solar arrays

  Floating solar arrays are PV systems that float on the surface of drinking water reservoirs, quarry lakes, irrigation canals or remediation and tailing ponds. A small number of such systems exist in France, India, Japan, South Korea, the United Kingdom, Singapore and the United States. The systems are said to have advantages over photovoltaics on land. The cost of land is more expensive, and there are fewer rules and regulations for structures built on bodies of water not used for recreation. Unlike most land-based solar plants, floating arrays can be unobtrusive because they are hidden from public view. They achieve higher efficiencies than PV panels on land, because water cools the panels. The panels have a special coating to prevent rust or corrosion. In May 2008, the Far Niente Winery in Oakville, California, pioneered the world's first floatovoltaic system by installing 994 solar PV modules with a total capacity of 477 kW onto 130 pontoons and floating them on the winery's irrigation pond. Utility-scale floating PV farms are starting to be built. Kyocera will develop the world's largest, a 13.4 MW farm on the reservoir above Yamakura Dam in Chiba Prefecture using 50,000 solar panels. Salt-water resistant floating farms are also being constructed for ocean use. The largest so far announced floatovoltaic project is a 350 MW power station in the Amazon region of Brazil.

- Solar-assisted heat pump

  A heat pump is a device that provides heat energy from a source of heat to a destination called a "heat sink". Heat pumps are designed to move thermal energy opposite to the direction of spontaneous heat flow by absorbing heat from a cold space and releasing it to a warmer one. A solar-assisted heat pump represents the integration of a heat pump and thermal solar panels in a single integrated system. Typically these two technologies are used separately (or only placing them in parallel) to produce hot water. In this system the solar thermal panel performs the function of the low temperature heat source and the heat produced is used to feed the heat pump's evaporator. The goal of this system is to get high COP and then produce energy in a more efficient and less expensive way.

  It is possible to use any type of solar thermal panel (sheet and tubes, roll-bond, heat pipe, thermal plates) or hybrid (mono/polycrystalline, thin film) in combination with the heat pump. The use of a hybrid panel is preferable because it allows to cover a part of the electricity demand of the heat pump and reduce the power consumption and consequently the variable costs of the system.

- Artificial photosynthesis

  Artificial photosynthesis uses techniques including nanotechnology to store solar electromagnetic energy in chemical bonds by splitting water to produce hydrogen and then using carbon dioxide to make methanol. Researchers in this field are striving to design molecular mimics of photosynthesis that utilize a wider region of the solar spectrum, employ catalytic systems made from abundant, inexpensive materials that are robust, readily repaired, non-toxic, stable in a variety of environmental conditions and perform more efficiently allowing a greater proportion of photon energy to end up in the storage compounds, i.e., carbohydrates (rather than building and sustaining living cells). However, prominent research faces hurdles, Sun Catalytix a MIT spin-off stopped scaling up their prototype fuel-cell in 2012, because it offers few savings over other ways to make hydrogen from sunlight.

- Algae fuels

  Producing liquid fuels from oil-rich varieties of algae is an ongoing research topic. Various microalgae grown in open or closed systems are being tried including some system that can be set up in brownfield and desert lands.

- Solar aircraft

  An electric aircraft is an aircraft that runs on electric motors rather than internal combustion engines, with electricity coming from fuel cells, solar cells, ultracapacitors, power beaming, or batteries.

In 2016, Solar Impulse 2 was the first solar-powered aircraft to complete circumnavigation of the world.

Currently, flying manned electric aircraft are mostly experimental demonstrators, though many small unmanned aerial vehicles are powered by batteries. Electrically powered model aircraft have been flown since the 1970s, with one report in 1957. The first man-carrying electrically powered flights were made in 1973. Between 2015-2016, a manned, solar-powered plane, Solar Impulse 2, completed a circumnavigation of the Earth.

- Solar updraft tower

  The Solar updraft tower is a renewable-energy power plant for generating electricity from low temperature solar heat. Sunshine heats the air beneath a very wide greenhouse-like roofed collector structure surrounding the central base of a very tall chimney tower. The resulting convection causes a hot air updraft in the tower by the chimney effect. This airflow drives wind turbines placed in the chimney updraft or around the chimney base to produce electricity. Plans for scaled-up versions of demonstration models will allow significant power generation, and may allow development of other applications, such as water extraction or distillation, and agriculture or horticulture. A more advanced version of a similarly themed technology is the Vortex engine which aims to replace large physical chimneys with a vortex of air created by a shorter, less-expensive structure.

- Space-based solar power

  For either photovoltaic or thermal systems, one option is to loft them into space, particularly Geosynchronous orbit. To be competitive with Earth-based solar power systems, the specific mass (kg/kW) times the cost to loft mass plus the cost of the parts needs to be $2400 or less. I.e., for a parts cost plus rectenna of $1100/kW, the product of the $/kg and kg/kW must be $1300/kW or less. Thus for 6.5 kg/kW, the transport cost cannot exceed $200/kg. While that will require a 100 to one reduction, SpaceX is targeting a ten to one reduction, Reaction Engines may make a 100 to one reduction possible.

- Carbon-neutral and negative fuels

  Carbon-neutral fuels are synthetic fuels (including methane, gasoline, diesel fuel, jet fuel or ammonia) produced by hydrogenating waste carbon dioxide recycled from power plant flue-gas emissions, recovered from automotive exhaust gas, or derived from carbonic acid in seawater. Such fuels are considered carbon-neutral because they do not result in a net increase in atmospheric greenhouse gases. T

  Carbon-neutral fuels offer relatively low cost energy storage, alleviating the problems of wind and solar variability, and they enable distribution of wind, water, and solar power through existing natural gas pipelines. Nighttime wind power is considered the most economical form of electrical power with which to synthesize fuel, because the load curve for electricity peaks sharply during the warmest hours of the day, but wind tends to blow slightly more at night than during the day, so, the price of nighttime wind power is often much less expensive than any alternative. Germany has built a 250 kilowatt synthetic methane plant which they are scaling up to 10 megawatts.

  The George Olah carbon dioxide recycling plant in Grindavík, Iceland has been producing 2 million liters of methanol transportation fuel per year from flue exhaust of the Svartsengi Power Station since 2011. It has the capacity to produce 5 million liters per year.

Debate

Renewable electricity production, from sources such as wind power and solar power, is sometimes criticized for being variable or intermittent, but is not true for concentrated solar, geothermal and biofuels, that have continuity. In any case, the International Energy Agency has stated that deployment of renewable technologies usually increases the diversity of electricity sources and, through local generation, contributes to the flexibility of the system and its resistance to central shocks.

There have been "not in my back yard" (NIMBY) concerns relating to the visual and other impacts of some wind farms, with local residents sometimes fighting or blocking construction. In the USA, the Massachusetts Cape Wind project was delayed for years partly because of aesthetic concerns. However, residents in other areas have been more positive. According to a town councilor, the overwhelming majority of locals believe that the Ardrossan Wind Farm in Scotland has enhanced the area.

A recent UK Government document states that "projects are generally more likely to succeed if they have broad public support and the consent of local communities. This means giving communities both a say and a stake". In countries such as Germany and Denmark many renewable projects are owned by communities, particularly through cooperative structures, and contribute significantly to overall levels of renewable energy deployment.

The market for renewable energy technologies has continued to grow. Climate change concerns and increasing in green jobs, coupled with high oil prices, peak oil, oil wars, oil spills, promotion of electric vehicles and renewable electricity, nuclear disasters and increasing government support, are driving increasing renewable energy legislation, incentives and commercialization. New government spending, regulation and policies helped the industry weather the 2009 economic crisis better than many other sectors.

Renewable Energy Commercialization

Renewable energy commercialization involves the deployment of three generations of renewable energy technologies dating back more than 100 years. First-generation technologies, which are already mature and economically competitive, include biomass, hydroelectricity, geothermal power and heat. Second-generation technologies are market-ready and are being deployed at the present time; they include solar heating, photovoltaics, wind power, solar thermal power stations, and modern forms of bioenergy.

Third-generation technologies require continued R&D efforts in order to make large contributions on a global scale and include advanced biomass gasification, hot-dry-rock geothermal power, and ocean energy. As of 2012, renewable energy accounts for about half of new nameplate electrical capacity installed and costs are continuing to fall.

The 150 MW Andasol solar power station is a commercial parabolic trough solar thermal power plant, located in Spain. The Andasol plant uses tanks of molten salt to store solar energy so that it can continue generating electricity even when the sun isn't shining.

Public policy and political leadership helps to "level the playing field" and drive the wider acceptance of renewable energy technologies. Countries such as Germany, Denmark, and Spain have led the way in implementing innovative policies which has driven most of the growth over the past decade. As of 2014, Germany has a commitment to the "Energiewende" transition to a sustainable energy economy, and Denmark has a commitment to 100% renewable energy by 2050. There are now 144 countries with renewable energy policy targets.

Renewable energy continued its rapid growth in 2015, providing multiple benefits. There was a new record set for installed wind and photovoltaic capacity (64GW and 57GW) and a new high of US$329 Billion for global renewables investment. A key benefit that this investment growth brings is a growth in jobs. The top countries for investment in recent years were China, Germany, Spain, the United States, Italy, and Brazil. Renewable energy companies include BrightSource Energy, First Solar, Gamesa, GE Energy, Goldwind, Sinovel, Trina Solar, Vestas, and Yingli.

Climate change concerns are also driving increasing growth in the renewable energy industries. According to a 2011 projection by the (IEA) International Energy Agency, solar power generators may produce most of the world's electricity within 50 years, reducing harmful greenhouse gas emissions.

Overview

Rationale for Renewables

Climate change, pollution, and energy insecurity are significant problems, and addressing them requires major changes to energy infrastructures. Renewable energy technologies are essential contributors to the energy supply portfolio, as they contribute to world energy security, reduce dependency on fossil fuels, and provide opportunities for mitigating greenhouse gases. Climate-disrupting fossil fuels are being replaced by clean, climate-stabilizing, non-depletable sources of energy:

...the transition from coal, oil, and gas to wind, solar, and geothermal energy is well under way. In the old economy, energy was produced by burning something — oil, coal, or natural gas — leading to the carbon emissions that have come to define our economy. The new energy economy harnesses the energy in wind, the energy coming from the sun, and heat from within the earth itself.

In international public opinion surveys there is strong support for a variety of methods for addressing the problem of energy supply. These methods include promoting renewable sources such as solar power and wind power, requiring utilities to use more renewable energy, and providing tax incentives to encourage the development and use of such technologies. It is expected that renewable energy investments will pay off economically in the long term.

EU member countries have shown support for ambitious renewable energy goals. In 2010, Eurobarometer polled the twenty-seven EU member states about the target "to increase the share of renewable energy in the EU by 20 percent by 2020". Most people in all twenty-seven countries either approved of the target or called for it to go further. Across the EU, 57 percent thought the proposed goal was "about right" and 16 percent thought it was "too modest." In comparison, 19 percent said it was "too ambitious".

As of 2011, new evidence has emerged that there are considerable risks associated with traditional energy sources, and that major changes to the mix of energy technologies is needed:

Several mining tragedies globally have underscored the human toll of the coal supply chain. New EPA initiatives targeting air toxics, coal ash, and effluent releases highlight the environmental impacts of coal and the cost of addressing them with control technologies. The use of fracking in natural gas exploration is coming under scrutiny, with evidence of groundwater contamination and greenhouse gas emissions. Concerns are increasing about the vast amounts of water used at coal-fired and nuclear power plants, particularly in regions of the country facing water shortages. Events at the Fukushima nuclear plant have renewed doubts about the ability to operate large numbers of nuclear plants safely over the long term. Further, cost estimates for "next generation" nuclear units continue to climb, and lenders are unwilling to finance these plants without taxpayer guarantees.

The 2014 REN21 Global Status Report says that renewable energies are no longer just energy sources, but ways to address pressing social, political, economic and environmental problems:

Today, renewables are seen not only as sources of energy, but also as tools to address many other pressing needs, including: improving energy security; reducing the health and environmental impacts associated with fossil and nuclear energy; mitigating greenhouse gas emissions; improving educational opportunities; creating jobs; reducing poverty; and increasing gender equality... Renewables have entered the mainstream.

Growth of Renewables

In 2008 for the first time, more renewable energy than conventional power capacity was added in both the European Union and United States, demonstrating a "fundamental transition" of the world's energy markets towards renewables, according to a report released by REN21, a global renewable energy policy network based in Paris. In 2010, renewable power consisted about a third of the newly built power generation capacities.

By the end of 2011, total renewable power capacity worldwide exceeded 1,360 GW, up 8%. Renewables producing electricity accounted for almost half of the 208 GW of capacity added globally during 2011. Wind and solar photovoltaics (PV) accounted for almost 40% and 30% . Based on REN21's 2014 report, renewables contributed 19 percent to our energy consumption and 22 percent to our electricity generation in 2012 and 2013, respectively. This energy consumption is divided as 9% coming from traditional biomass, 4.2% as heat energy (non-biomass), 3.8% hydro electricity and 2% electricity from wind, solar, geothermal, and biomass.

During the five-years from the end of 2004 through 2009, worldwide renewable energy capacity grew at rates of 10–60 percent annually for many technologies. In 2011, UN under-secretary general Achim Steiner said: "The continuing growth in this core segment of the green economy is not happening by chance. The combination of government target-setting, policy support and stimulus funds is underpinning the renewable industry's rise and bringing the much needed transformation of our global energy system within reach." He added: "Renewable energies are expanding both in terms of investment, projects and geographical spread. In doing so, they are making an increasing contribution to combating climate change, countering energy poverty and energy insecurity".

According to a 2011 projection by the International Energy Agency, solar power plants may produce most of the world's electricity within 50 years, significantly reducing the emissions of greenhouse gases that harm the environment. The IEA has said: "Photovoltaic and solar-thermal plants may meet most of the world's demand for electricity by 2060 – and half of all energy needs – with wind, hydropower and biomass plants supplying much of the remaining generation". "Photovoltaic and concentrated solar power together can become the major source of electricity".

Selected renewable energy indicators													
Selected global indicators	**2004**	**2005**	**2006**	**2007**	**2008**	**2009**	**2010**	**2011**	**2012**	**2013**	**2014**	**2015**	**units**
Investment in new renewable capacity (annual)	30	38	63	104	130	160	211	257	244	214	270	285	billion USD
Existing renewables power capacity, including large-scale hydro	895	930	1,020	1,070	1,140	1,230	1,320	1,360	1,470	1,560	1,712	1,849	GWe
Existing renewables power capacity, ex-cluding large hydro					200	250	312	390	480	560	657	785	GWe
Hydropower capacity (existing)						915	945	970	990	1,000	1,055	1,064	GWe
Wind power capacity (existing)	48	59	74	94	121	159	198	238	283	318	370	433	GWe
Solar PV capacity (grid-connected)				7.6	16	23	40	70	100	139	177	227	GWe
Solar hot water capacity (existing)	77	88	105	120	130	160	185	232	255	326	406	435	GWth
Ethanol production (annual)	30.5	33	39	50	67	76	86	86	83	87	94	98	billion liters
Biodiesel production (annual)					12	17	19	21	22	26	29.7	30	billion liters
Countries with policy targets for renew-able energy use	45	49		68	79	89	98	118	138	144	164	173	

In 2013, China led the world in renewable energy production, with a total capacity of 378 GW, mainly from hydroelectric and wind power. As of 2014, China leads the world in the production and use of wind power, solar photovoltaic power and smart grid technologies, generating almost as much water, wind and solar energy as all of France and Germany's power plants combined. China's renewable energy sector is growing faster than its fossil fuels and nuclear power capacity. Since 2005, production of solar cells in China has expanded 100-fold. As Chinese renewable manufacturing has grown, the costs of renewable energy technologies have dropped. Innovation has helped, but the main driver of reduced costs has been market expansion.

Economic Trends

Renewable energy technologies are getting cheaper, through technological change and through the benefits of mass production and market competition. A 2011 IEA report said: "A portfolio of renewable energy technologies is becoming cost-competitive in an increasingly broad range of circumstances, in some cases providing investment opportunities without the need for specific economic support," and added that "cost reductions in critical technologies, such as wind and solar, are set to continue." As of 2011, there have been substantial reductions in the cost of solar and wind technologies:

The price of PV modules per MW has fallen by 60 percent since the summer of 2008, according to Bloomberg New Energy Finance estimates, putting solar power for the first time on a competitive footing with the retail price of electricity in a number of sunny countries. Wind turbine prices have also fallen – by 18 percent per MW in the last two years – reflecting, as with solar, fierce competition in the supply chain. Further improvements in the levelised cost of energy for solar, wind and other technologies lie ahead, posing a growing threat to the dominance of fossil fuel generation sources in the next few years.

Hydro-electricity and geothermal electricity produced at favourable sites are now the cheapest way to generate electricity. Renewable energy costs continue to drop, and the levelised cost of electricity (LCOE) is declining for wind power, solar photovoltaic (PV), concentrated solar power (CSP) and some biomass technologies.

Renewable energy is also the most economic solution for new grid-connected capacity in areas with good resources. As the cost of renewable power falls, the scope of economically viable applications increases. Renewable technologies are now often the most economic solution for new generating capacity. Where "oil-fired generation is the predominant power generation source (e.g. on islands, off-grid and in some countries) a lower-cost renewable solution almost always exists today". As of 2012, renewable power generation technologies accounted for around half of all new power generation capacity additions globally. In 2011, additions included 41 gigawatt (GW) of new wind power capacity, 30 GW of PV, 25 GW of hydro-electricity, 6 GW of biomass, 0.5 GW of CSP, and 0.1 GW of geothermal power.

Three Generations of Technologies

Renewable energy includes a number of sources and technologies at different stages of commercialization. The International Energy Agency (IEA) has defined three generations of renewable energy technologies, reaching back over 100 years:

- "*First-generation technologies* emerged from the industrial revolution at the end of the 19th century and include hydropower, biomass combustion, geothermal power and heat. These technologies are quite widely used.
- *Second-generation technologies* include solar heating and cooling, wind power, modern forms of bioenergy, and solar photovoltaics. These are now entering markets as a result of research, development and demonstration (RD&D) investments since the 1980s. Initial investment was prompted by energy security concerns linked to the oil crises of the 1970s but the enduring appeal of these technologies is due, at least in part, to environmental benefits. Many of the technologies reflect significant advancements in materials.

- *Third-generation technologies* are still under development and include advanced biomass gasification, biorefinery technologies, concentrating solar thermal power, hot-dry-rock geothermal power, and ocean energy. Advances in nanotechnology may also play a major role". First-generation technologies are well established, second-generation technologies are entering markets, and third-generation technologies heavily depend on long-term research and development commitments, where the public sector has a role to play.

First-generation Technologies

Biomass heating plant in Austria. The total heat power is about 1000 kW.

First-generation technologies are widely used in locations with abundant resources. Their future use depends on the exploration of the remaining resource potential, particularly in developing countries, and on overcoming challenges related to the environment and social acceptance.

Biomass

Biomass for heat and power is a fully mature technology which offers a ready disposal mechanism for municipal, agricultural, and industrial organic wastes. However, the industry has remained relatively stagnant over the decade to 2007, even though demand for biomass (mostly wood) continues to grow in many developing countries. One of the problems of biomass is that material directly combusted in cook stoves produces pollutants, leading to severe health and environmental consequences, although improved cook stove programmes are alleviating some of these effects. First-generation biomass technologies can be economically competitive, but may still require deployment support to overcome public acceptance and small-scale issues.

Hydroelectricity

Hydroelectricity is the term referring to electricity generated by hydropower; the production of electrical power through the use of the gravitational force of falling or flowing water. In 2015 hydropower generated 16.6% of the worlds total electricity and 70% of all renewable electricity and is expected to increase about 3.1% each year for the next 25 years. Hydroelectric plants have the advantage of being long-lived and many existing plants have operated for more than 100 years.

Hydropower is produced in 150 countries, with the Asia-Pacific region generating 32 percent of global hydropower in 2010. China is the largest hydroelectricity producer, with 721 terawatt-hours of production in 2010, representing around 17 percent of domestic electricity use. There are now three hydroelectricity plants larger than 10 GW: the Three Gorges Dam in China, Itaipu Dam across the Brazil/Paraguay border, and Guri Dam in Venezuela. The cost of hydroelectricity is low, making it a competitive source of renewable electricity. The average cost of electricity from a hydro plant larger than 10 megawatts is 3 to 5 U.S. cents per kilowatt-hour.

Geothermal Power and Heat

Geothermal power plants can operate 24 hours per day, providing baseload capacity. Estimates for the world potential capacity for geothermal power generation vary widely, ranging from 40 GW by 2020 to as much as 6,000 GW.

Geothermal power capacity grew from around 1 GW in 1975 to almost 10 GW in 2008. The United States is the world leader in terms of installed capacity, representing 3.1 GW. Other countries with significant installed capacity include the Philippines (1.9 GW), Indonesia (1.2 GW), Mexico (1.0 GW), Italy (0.8 GW), Iceland (0.6 GW), Japan (0.5 GW), and New Zealand (0.5 GW). In some countries, geothermal power accounts for a significant share of the total electricity supply, such as in the Philippines, where geothermal represented 17 percent of the total power mix at the end of 2008.

Geothermal (ground source) heat pumps represented an estimated 30 GWth of installed capacity at the end of 2008, with other direct uses of geothermal heat (i.e., for space heating, agricultural drying and other uses) reaching an estimated 15 GWth. As of 2008, at least 76 countries use direct geothermal energy in some form.

Second-generation Technologies

Markets for second-generation technologies have been strong and growing over the past decade, and these technologies have gone from being a passion for the dedicated few to a major economic sector in countries such as Germany, Spain, the United States, and Japan. Many large industrial companies and financial institutions are involved and the challenge is to broaden the market base for continued growth worldwide.

Solar Heating

Solar energy technologies, such as solar water heaters, located on or near the buildings which they supply with energy, are a prime example of a soft energy technology.

Solar heating systems are a well known second-generation technology and generally consist of solar thermal collectors, a fluid system to move the heat from the collector to its point of usage, and a reservoir or tank for heat storage. The systems may be used to heat domestic hot water, swimming pools, or homes and businesses. The heat can also be used for industrial process applications or as an energy input for other uses such as cooling equipment.

In many warmer climates, a solar heating system can provide a very high percentage (50 to 75%) of domestic hot water energy. As of 2009, China has 27 million rooftop solar water heaters.

Photovoltaics

Nellis Solar Power Plant at Nellis Air Force Base. These panels track the sun in one axis.

President Barack Obama speaks at the DeSoto Next Generation Solar Energy Center.

Photovoltaic (PV) cells, also called solar cells, convert light into electricity. In the 1980s and early 1990s, most photovoltaic modules were used to provide remote-area power supply, but from around 1995, industry efforts have focused increasingly on developing building integrated photovoltaics and photovoltaic power stations for grid connected applications.

Many solar photovoltaic power stations have been built, mainly in Europe. As of July 2012, the largest photovoltaic (PV) power plants in the world are the Agua Caliente Solar Project (USA, 247 MW), Charanka Solar Park (India, 214 MW), Golmud Solar Park (China, 200 MW), Perovo Solar Park (Russia 100 MW), Sarnia Photovoltaic Power Plant (Canada, 97 MW), Brandenburg-Briest Solarpark (Germany 91 MW), Solarpark Finow Tower (Germany 84.7 MW), Montalto di Castro Photovoltaic Power Station (Italy, 84.2 MW), Eggebek Solar Park (Germany 83.6 MW), Senftenberg Solarpark (Germany 82 MW), Finsterwalde Solar Park (Germany, 80.7 MW), Okhotnykovo Solar Park (Russia, 80 MW), Lopburi Solar Farm (Thailand 73.16 MW), Rovigo Photovoltaic Power Plant (Italy, 72 MW), and the Lieberose Photovoltaic Park (Germany, 71.8 MW).

There are also many large plants under construction. The Desert Sunlight Solar Farm under construction in Riverside County, California and Topaz Solar Farm being built in San Luis Obispo County, California are both 550 MW solar parks that will use thin-film solar photovoltaic modules made by First Solar. The Blythe Solar Power Project is a 500 MW photovoltaic station under construction in Riverside County, California. The California Valley Solar Ranch (CVSR) is a 250 megawatt (MW) solar photovoltaic power plant, which is being built by SunPower in the Carrizo Plain, northeast of California Valley. The 230 MW Antelope Valley Solar Ranch is a First Solar photovoltaic project which is under construction in the Antelope Valley area of the Western Mojave Desert, and due to be completed in 2013. The Mesquite Solar project is a photovoltaic solar power plant being built in Arlington, Maricopa County, Arizona, owned by Sempra Generation. Phase 1 will have a nameplate capacity of 150 megawatts.

Many of these plants are integrated with agriculture and some use innovative tracking systems that follow the sun's daily path across the sky to generate more electricity than conventional fixed-mounted systems. There are no fuel costs or emissions during operation of the power stations.

Wind Power

Some of the second-generation renewables, such as wind power, have high potential and have already realised relatively low production costs. Global wind power installations increased by 35,800 MW in 2010, bringing total installed capacity up to 194,400 MW, a 22.5% increase on the 158,700 MW installed at the end of 2009. The increase for 2010 represents investments totalling €47.3 billion (US$65 billion) and for the first time more than half of all new wind power was added outside of the traditional markets of Europe and North America, mainly driven, by the continuing boom in China which accounted for nearly half of all of the installations at 16,500 MW. China now has 42,300 MW of wind power installed. Wind power accounts for approximately 19% of electricity generated in Denmark, 9% in Spain and Portugal, and 6% in Germany and the Republic of Ireland. In Australian state of South Australia wind power, championed by Premier Mike Rann (2002–2011), now comprises 26% of the state's electricity generation, edging out coal fired power. At the end of 2011 South Australia, with 7.2% of Australia's population, had 54%of the nation's installed wind power capacity. Wind power's share of worldwide electricity usage at the end of 2014 was 3.1%. These are some of the largest wind farms in the world:

Landowners in the US typically receive $3,000 to $5,000 per year in rental income from each wind turbine, while farmers continue to grow crops or graze cattle up to the foot of the turbines.

Large onshore wind farms		
Wind farm	**Current capacity (MW)**	**Country**
Gansu Wind Farm	6,000	China
Alta (Oak Creek-Mojave)	1,320	United States
Jaisalmer Wind Park	1,064	India
Shepherds Flat Wind Farm	845	United States
Roscoe Wind Farm	782	United States
Horse Hollow Wind Energy Center	736	United States
Capricorn Ridge Wind Farm	662	United States
Fântânele-Cogealac Wind Farm	600	Romania
Fowler Ridge Wind Farm	600	United States
Whitelee Wind Farm	539	United Kingdom

As of 2014, the wind industry in the USA is able to produce more power at lower cost by using taller wind turbines with longer blades, capturing the faster winds at higher elevations. This has opened up new opportunities and in Indiana, Michigan, and Ohio, the price of power from wind turbines built 300 feet to 400 feet above the ground can now compete with conventional fossil fuels like coal. Prices have fallen to about 4 cents per kilowatt-hour in some cases and utilities have been increasing the amount of wind energy in their portfolio, saying it is their cheapest option.

Solar Thermal Power Stations

View of Ivanpah Solar Electric Generating System from Yates Well Road, San Bernardino County, California. The Clark Mountain Range can be seen in the distance.

Solar thermal power stations include the 354 megawatt (MW) Solar Energy Generating Systems power plant in the USA, Solnova Solar Power Station (Spain, 150 MW), Andasol solar power station (Spain, 100 MW), Nevada Solar One (USA, 64 MW), PS20 solar power tower (Spain, 20 MW), and the PS10 solar power tower (Spain, 11 MW). The 370 MW Ivanpah Solar Power Facility, located in California's Mojave Desert, is the world's largest solar-thermal power plant project currently under construction. Many other plants are under construction or planned, mainly in Spain and the USA. In developing countries, three World Bank projects for integrated solar thermal/combined-cycle gas-turbine power plants in Egypt, Mexico, and Morocco have been approved.

Modern forms of Bioenergy

Global ethanol production for transport fuel tripled between 2000 and 2007 from 17 billion to more than 52 billion litres, while biodiesel expanded more than tenfold from less than 1 billion to almost 11 billion litres. Biofuels provide 1.8% of the world's transport fuel and recent estimates indicate a continued high growth. The main producing countries for transport biofuels are the USA, Brazil, and the EU.

Brazil has one of the largest renewable energy programs in the world, involving production of ethanol fuel from sugar cane, and ethanol now provides 18 percent of the country's automotive fuel. As a result of this and the exploitation of domestic deep water oil sources, Brazil, which for years had to import a large share of the petroleum needed for domestic consumption, recently reached complete self-sufficiency in liquid fuels.

Nearly all the gasoline sold in the United States today is mixed with 10 percent ethanol, a mix known as E10, and motor vehicle manufacturers already produce vehicles designed to run on much higher ethanol blends. Ford, DaimlerChrysler, and GM are among the automobile companies that sell flexible-fuel cars, trucks, and minivans that can use gasoline and ethanol blends ranging from pure gasoline up to 85% ethanol (E85). The challenge is to expand the market for biofuels beyond the farm states where they have been most popular to date. The Energy Policy Act of 2005, which calls for 7.5 billion US gallons (28,000,000 m^3) of biofuels to be used annually by 2012, will also help to expand the market.

The growing ethanol and biodiesel industries are providing jobs in plant construction, operations, and maintenance, mostly in rural communities. According to the Renewable Fuels Association, "the ethanol industry created almost 154,000 U.S. jobs in 2005 alone, boosting household income by $5.7 billion. It also contributed about $3.5 billion in tax revenues at the local, state, and federal levels".

Third-generation Technologies

Third-generation renewable energy technologies are still under development and include advanced biomass gasification, biorefinery technologies, hot-dry-rock geothermal power, and ocean energy. Third-generation technologies are not yet widely demonstrated or have limited commercialization. Many are on the horizon and may have potential comparable to other renewable energy technologies, but still depend on attracting sufficient attention and research and development funding.

New Bioenergy Technologies

Selected Commercial Cellulosic Ethanol Plants in the U.S.		
Company	**Location**	**Feedstock**
Abengoa Bioenergy	Hugoton, KS	Wheat straw
BlueFire Ethanol	Irvine, CA	Multiple sources
Gulf Coast Energy	Mossy Head, FL	Wood waste
Mascoma	Lansing, MI	Wood
POET LLC	Emmetsburg, IA	Corn cobs
SunOpta	Little Falls, MN	Wood chips
Xethanol	Auburndale, FL	Citrus peels
Note: plants are either operational or under construction		

According to the International Energy Agency, cellulosic ethanol biorefineries could allow biofuels to play a much bigger role in the future than organizations such as the

IEA previously thought. Cellulosic ethanol can be made from plant matter composed primarily of inedible cellulose fibers that form the stems and branches of most plants. Crop residues (such as corn stalks, wheat straw and rice straw), wood waste, and municipal solid waste are potential sources of cellulosic biomass. Dedicated energy crops, such as switchgrass, are also promising cellulose sources that can be sustainably produced in many regions.

Ocean Energy

Ocean energy is all forms of renewable energy derived from the sea including wave energy, tidal energy, river current, ocean current energy, offshore wind, salinity gradient energy and ocean thermal gradient energy.

The Rance Tidal Power Station (240 MW) is the world's first tidal power station. The facility is located on the estuary of the Rance River, in Brittany, France. Opened on 26 November 1966, it is currently operated by Électricité de France, and is the largest tidal power station in the world, in terms of installed capacity.

First proposed more than thirty years ago, systems to harvest utility-scale electrical power from ocean waves have recently been gaining momentum as a viable technology. The potential for this technology is considered promising, especially on west-facing coasts with latitudes between 40 and 60 degrees:

In the United Kingdom, for example, the Carbon Trust recently estimated the extent of the economically viable offshore resource at 55 TWh per year, about 14% of current national demand. Across Europe, the technologically achievable resource has been estimated to be at least 280 TWh per year. In 2003, the U.S. Electric Power Research Institute (EPRI) estimated the viable resource in the United States at 255 TWh per year (6% of demand).

There are currently nine projects, completed or in-development, off the coasts of the United Kingdom, United States, Spain and Australia to harness the rise and fall of waves by Ocean Power Technologies. The current maximum power output is 1.5 MW (Reedsport, Oregon), with development underway for 100 MW (Coos Bay, Oregon).

Enhanced Geothermal Systems

As of 2008, geothermal power development was under way in more than 40 countries, partially attributable to the development of new technologies, such as Enhanced Geothermal Systems. The development of binary cycle power plants and improvements in drilling and extraction technology may enable enhanced geothermal systems over a much greater geographical range than "traditional" Geothermal systems. Demonstration EGS projects are operational in the USA, Australia, Germany, France, and The United Kingdom.

Advanced Solar Concepts

Beyond the already established solar photovoltaics and solar thermal power technologies are such advanced solar concepts as the solar updraft tower or space-based solar power. These concepts have yet to (if ever) be commercialized.

The Solar updraft tower (SUT) is a renewable-energy power plant for generating electricity from low temperature solar heat. Sunshine heats the air beneath a very wide greenhouse-like roofed collector structure surrounding the central base of a very tall chimney tower. The resulting convection causes a hot air updraft in the tower by the chimney effect. This airflow drives wind turbines placed in the chimney updraft or around the chimney base to produce electricity. Plans for scaled-up versions of demonstration models will allow significant power generation, and may allow development of other applications, such as water extraction or distillation, and agriculture or horticulture.

A more advanced version of a similarly themed technology is the Vortex engine (AVE) which aims to replace large physical chimneys with a vortex of air created by a shorter, less-expensive structure.

Space-based solar power (SBSP) is the concept of collecting solar power in space (using an "SPS", that is, a "solar-power satellite" or a "satellite power system") for use on Earth. It has been in research since the early 1970s. SBSP would differ from current solar collection methods in that the means used to collect energy would reside on an orbiting satellite instead of on Earth's surface. Some projected benefits of such a system are a higher collection rate and a longer collection period due to the lack of a diffusing atmosphere and night time in space.

Renewable Energy Industry

A Vestas wind turbine

Monocrystalline solar cell

Total investment in renewable energy reached $211 billion in 2010, up from $160 billion in 2009. The top countries for investment in 2010 were China, Germany, the United States, Italy, and Brazil. Continued growth for the renewable energy sector is expected and promotional policies helped the industry weather the 2009 economic crisis better than many other sectors.

Wind Power Companies

As of 2010, Vestas (from Denmark) is the world's top wind turbine manufacturer in terms of percentage of market volume, and Sinovel (from China) is in second place. Together Vestas and Sinovel delivered 10,228 MW of new wind power capacity in 2010, and their market share was 25.9 percent. GE Energy (USA) was in third place, closely followed by Goldwind, another Chinese supplier. German Enercon ranks fifth in the world, and is followed in sixth place by Indian-based Suzlon.

Photovoltaic Market Trends

The solar PV market has been growing for the past few years. According to solar PV research company, PVinsights, worldwide shipment of solar modules in 2011 was around 25 GW, and the shipment year over year growth was around 40%. The top 5 solar module players in 2011 in turns are Suntech, First Solar, Yingli, Trina, and Sungen. The top 5 solar module companies possessed 51.3% market share of solar modules, according to PVinsights' market intelligence report.

2013 Ranking	Solar Module Company	Change from 2012	Country
1	Yingli Green Energy	–	China

2	Trina Solar	+1	China
3	Sharp Solar	+3	Japan
4	Canadian Solar	–	Canada
5	Jinko Solar	+3	China
6	ReneSola	+7	China
7	First Solar	−2	China
8	Hanwha Solarone	+2	South Korea
9	Kyocera	+5	Japan
10	JA Solar	−3	China
Sources:			

The PV industry has seen drops in module prices since 2008. In late 2011, factory-gate prices for crystalline-silicon photovoltaic modules dropped below the $1.00/W mark. The $1.00/W installed cost, is often regarded in the PV industry as marking the achievement of grid parity for PV. These reductions have taken many stakeholders, including industry analysts, by surprise, and perceptions of current solar power economics often lags behind reality. Some stakeholders still have the perspective that solar PV remains too costly on an unsubsidized basis to compete with conventional generation options. Yet technological advancements, manufacturing process improvements, and industry re-structuring, mean that further price reductions are likely in coming years.

Top 10 PV countries in 2014 (MW)

Total capacity			**Added capacity**		
1.	Germany	38,200	1.	China	10,560
2.	China	28,199	2.	Japan	9,700
3.	Japan	23,300	3.	United States	6,201
4.	Italy	18,460	4.	UK	2,273
5.	United States	18,280	5.	Germany	1,900
6.	France	5,660	6.	France	927
7.	Spain	5,358	7.	Australia	910
8.	UK	5,104	8.	South Korea	909
9.	Australia	4,136	9.	South Africa	800
10.	Belgium	3,074	10.	India	616

Data: IEA-PVPS Snapshot of Global PV 1992–2014 report, March 2015

Non-technical Barriers to Acceptance

Many energy markets, institutions, and policies have been developed to support the production and use of fossil fuels. Newer and cleaner technologies may offer social and environmental benefits, but utility operators often reject renewable resources because they are trained to think only in terms of big, conventional power plants. Consumers often ignore renewable power systems because they are not given accurate price signals about electricity consumption. Intentional market distortions (such as subsidies), and unintentional market distortions (such as split incentives) may work against renewables. Benjamin K. Sovacool has argued that "some of the most surreptitious, yet powerful, impediments facing renewable energy and energy efficiency in the United States are more about *culture* and *institutions* than engineering and science".

The obstacles to the widespread commercialization of renewable energy technologies are primarily political, not technical, and there have been many studies which have identified a range of "non-technical barriers" to renewable energy use. These barriers are impediments which put renewable energy at a marketing, institutional, or policy disadvantage relative to other forms of energy. Key barriers include:

- Difficulty overcoming established energy systems, which includes difficulty introducing innovative energy systems, particularly for distributed generation such as photovoltaics, because of technological lock-in, electricity markets designed for centralized power plants, and market control by established operators. As the Stern Review on the Economics of Climate Change points out:

  National grids are usually tailored towards the operation of centralised power plants and thus favour their performance. Technologies that do not easily fit into these networks may struggle to enter the market, even if the technology itself is commercially viable. This applies to distributed generation as most grids are not suited to receive electricity from many small sources. Large-scale renewables may also encounter problems if they are sited in areas far from existing grids.

- Lack of government policy support, which includes the lack of policies and regulations supporting deployment of renewable energy technologies and the presence of policies and regulations hindering renewable energy development and supporting conventional energy development. Examples include subsidies for fossil-fuels, insufficient consumer-based renewable energy incentives, government underwriting for nuclear plant accidents, and complex zoning and permitting processes for renewable energy.

- Lack of information dissemination and consumer awareness.

- Higher capital cost of renewable energy technologies compared with conventional energy technologies.

- Inadequate financing options for renewable energy projects, including insufficient access to affordable financing for project developers, entrepreneurs and consumers.
- Imperfect capital markets, which includes failure to internalize all costs of conventional energy (e.g., effects of air pollution, risk of supply disruption) and failure to internalize all benefits of renewable energy (e.g., cleaner air, energy security).
- Inadequate workforce skills and training, which includes lack of adequate scientific, technical, and manufacturing skills required for renewable energy production; lack of reliable installation, maintenance, and inspection services; and failure of the educational system to provide adequate training in new technologies.
- Lack of adequate codes, standards, utility interconnection, and net-metering guidelines.
- Poor public perception of renewable energy system aesthetics.
- Lack of stakeholder/community participation and co-operation in energy choices and renewable energy projects.

With such a wide range of non-technical barriers, there is no "silver bullet" solution to drive the transition to renewable energy. So ideally there is a need for several different types of policy instruments to complement each other and overcome different types of barriers.

A policy framework must be created that will level the playing field and redress the imbalance of traditional approaches associated with fossil fuels. The policy landscape must keep pace with broad trends within the energy sector, as well as reflecting specific social, economic and environmental priorities.

Public Policy Landscape

Public policy has a role to play in renewable energy commercialization because the free market system has some fundamental limitations. As the Stern Review points out:

In a liberalised energy market, investors, operators and consumers should face the full cost of their decisions. But this is not the case in many economies or energy sectors. Many policies distort the market in favour of existing fossil fuel technologies.

The International Solar Energy Society has stated that "historical incentives for the conventional energy resources continue even today to bias markets by burying many of the real societal costs of their use".

Fossil-fuel energy systems have different production, transmission, and end-use costs and characteristics than do renewable energy systems, and new promotional policies are needed to ensure that renewable systems develop as quickly and broadly as is socially desirable.

Lester Brown states that the market "does not incorporate the indirect costs of providing goods or services into prices, it does not value nature's services adequately, and it does not respect the sustainable-yield thresholds of natural systems". It also favors the near term over the long term, thereby showing limited concern for future generations. Tax and subsidy shifting can help overcome these problems, though is also problematic to combine different international normative regimes regulating this issue.

Shifting Taxes

Tax shifting has been widely discussed and endorsed by economists. It involves lowering income taxes while raising levies on environmentally destructive activities, in order to create a more responsive market. For example, a tax on coal that included the increased health care costs associated with breathing polluted air, the costs of acid rain damage, and the costs of climate disruption would encourage investment in renewable technologies. Several Western European countries are already shifting taxes in a process known there as environmental tax reform.

In 2001, Sweden launched a new 10-year environmental tax shift designed to convert 30 billion kroner ($3.9 billion) of income taxes to taxes on environmentally destructive activities. Other European countries with significant tax reform efforts are France, Italy, Norway, Spain, and the United Kingdom. Asia's two leading economies, Japan and China, are considering carbon taxes.

Shifting Subsidies

Just as there is a need for tax shifting, there is also a need for subsidy shifting. Subsidies are not an inherently bad thing as many technologies and industries emerged through government subsidy schemes. The Stern Review explains that of 20 key innovations from the past 30 years, only one of the 14 was funded entirely by the private sector and nine were totally publicly funded. In terms of specific examples, the Internet was the result of publicly funded links among computers in government laboratories and research institutes. And the combination of the federal tax deduction and a robust state tax deduction in California helped to create the modern wind power industry.

Lester Brown has argued that "a world facing the prospect of economically disruptive climate change can no longer justify subsidies to expand the burning of coal and oil. Shifting these subsidies to the development of climate-benign energy sources such as wind, solar, biomass, and geothermal power is the key to stabilizing the earth's climate." The

International Solar Energy Society advocates "leveling the playing field" by redressing the continuing inequities in public subsidies of energy technologies and R&D, in which the fossil fuel and nuclear power receive the largest share of financial support.

Some countries are eliminating or reducing climate disrupting subsidies and Belgium, France, and Japan have phased out all subsidies for coal. Germany is reducing its coal subsidy. The subsidy dropped from $5.4 billion in 1989 to $2.8 billion in 2002, and in the process Germany lowered its coal use by 46 percent. China cut its coal subsidy from $750 million in 1993 to $240 million in 1995 and more recently has imposed a high-sulfur coal tax. However, the United States has been increasing its support for the fossil fuel and nuclear industries.

In November 2011, an IEA report entitled *Deploying Renewables 2011* said "subsidies in green energy technologies that were not yet competitive are justified in order to give an incentive to investing into technologies with clear environmental and energy security benefits". The IEA's report disagreed with claims that renewable energy technologies are only viable through costly subsidies and not able to produce energy reliably to meet demand.

A fair and efficient imposition of subsidies for renewable energies and aiming at sustainable development, however, require coordination and regulation at a global level, as subsidies granted in one country can easily disrupt industries and policies of others, thus underlining the relevance of this issue at the World Trade Organization.

Renewable Energy Targets

Setting national renewable energy targets can be an important part of a renewable energy policy and these targets are usually defined as a percentage of the primary energy and/or electricity generation mix. For example, the European Union has prescribed an indicative renewable energy target of 12 per cent of the total EU energy mix and 22 per cent of electricity consumption by 2010. National targets for individual EU Member States have also been set to meet the overall target. Other developed countries with defined national or regional targets include Australia, Canada, Israel, Japan, Korea, New Zealand, Norway, Singapore, Switzerland, and some US States.

National targets are also an important component of renewable energy strategies in some developing countries. Developing countries with renewable energy targets include China, India, Indonesia, Malaysia, the Philippines, Thailand, Brazil, Egypt, Mali, and South Africa. The targets set by many developing countries are quite modest when compared with those in some industrialized countries.

Renewable energy targets in most countries are indicative and nonbinding but they have assisted government actions and regulatory frameworks. The United Nations Environment Program has suggested that making renewable energy targets legally binding could be an important policy tool to achieve higher renewable energy market penetration.

Levelling the Playing Field

The IEA has identified three actions which will allow renewable energy and other clean energy technologies to "more effectively compete for private sector capital".

- "First, energy prices must appropriately reflect the "true cost" of energy (e.g. through carbon pricing) so that the positive and negative impacts of energy production and consumption are fully taken into account". Example: New UK nuclear plants cost £92.50/MWh, whereas offshore wind farms in the UK are supported with €74.2/MWh at a price of £150 in 2011 falling to £130 per MWh in 2022. In Denmark, the price can be €84/MWh.
- "Second, inefficient fossil fuel subsidies must be removed, while ensuring that all citizens have access to affordable energy".
- "Third, governments must develop policy frameworks that encourage private sector investment in lower-carbon energy options".

Green Stimulus Programs

In response to the global financial crisis in the late 2000s, the world's major governments made "green stimulus" programs one of their main policy instruments for supporting economic recovery. Some US$188 billion in green stimulus funding had been allocated to renewable energy and energy efficiency, to be spent mainly in 2010 and in 2011.

Energy Sector Regulation

Public policy determines the extent to which renewable energy (RE) is to be incorporated into a developed or developing country's generation mix. Energy sector regulators implement that policy—thus affecting the pace and pattern of RE investments and connections to the grid. Energy regulators often have authority to carry out a number of functions that have implications for the financial feasibility of renewable energy projects. Such functions include issuing licenses, setting performance standards, monitoring the performance of regulated firms, determining the price level and structure of tariffs, establishing uniform systems of accounts, arbitrating stakeholder disputes (like interconnection cost allocations), performing management audits, developing agency human resources (expertise), reporting sector and commission activities to government authorities, and coordinating decisions with other government agencies. Thus, regulators make a wide range of decisions that affect the financial outcomes associated with RE investments. In addition, the sector regulator is in a position to give advice to the government regarding the full implications of focusing on climate change or energy security. The energy sector regulator is the natural advocate for efficiency and cost-containment throughout the process of designing and implementing RE policies. Since policies are not self-implementing, energy sector regulators become a key facilitator (or blocker) of renewable energy investments.

Energy Transition in Germany

Photovoltaic array and wind turbines at the Schneebergerhof wind farm in the German state of Rheinland-Pfalz

The *Energiewende* (German for *energy transition*) is the transition by Germany to a low carbon, environmentally sound, reliable, and affordable energy supply. The new system will rely heavily on renewable energy (particularly wind, photovoltaics, and biomass) energy efficiency, and energy demand management. Most if not all existing coal-fired generation will need to be retired. The phase-out of Germany's fleet of nuclear reactors, to be complete by 2022, is a key part of the program.

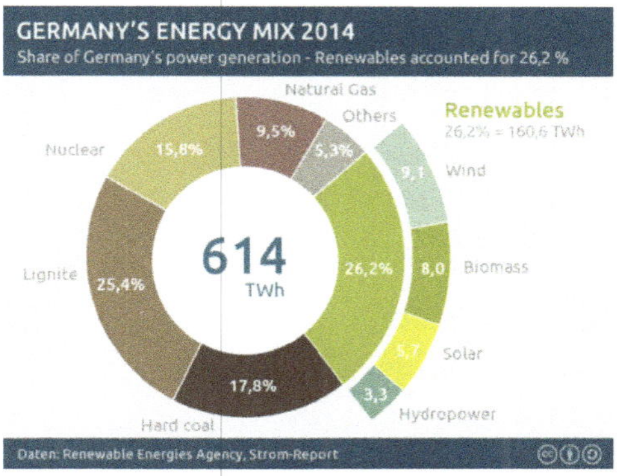

Market share of Germany's power generation 2014

Legislative support for the *Energiewende* was passed in late 2010 and includes greenhouse gas (GHG) reductions of 80–95% by 2050 (relative to 1990) and a renewable energy target of 60% by 2050. These targets are ambitious. One Berlin policy institute noted that "while the German approach is not unique worldwide, the speed and scope of the *Energiewende* are exceptional". The *Energiewende* also seeks a greater transparency in relation to national energy policy formation.

Germany has made significant progress on its GHG emissions reduction target, achieving a 27% decrease between 1990 and 2014. However Germany will need to maintain an average GHG emissions abatement rate of 3.5% per annum to reach its *Energiewende* goal, equal to the maximum historical value thus far.

Germany spends €1.5 billion per annum on energy research (2013 figure) in an effort to solve the technical and social issues raised by the transition. This includes a number of computer studies that have confirmed the feasibility and a similar cost (relative to business-as-usual and given that carbon is adequately priced) of the *Energiewende.*

These initiatives go well beyond European Union legislation and the national policies of other European states. The policy objectives have been embraced by the German federal government and has resulted in a huge expansion of renewables, particularly wind power. Germany's share of renewables has increased from around 5% in 1999 to 22.9% in 2012, surpassing the OECD average of 18% usage of renewables. Producers have been guaranteed a fixed feed-in tariff for 20 years, guaranteeing a fixed income. Energy co-operatives have been created, and efforts were made to decentralize control and profits. The large energy companies have a disproportionately small share of the renewables market. However, in some cases poor investment designs have caused bankruptcies and low returns, and unrealistic promises have been shown to be far from reality. Nuclear power plants were closed, and the existing nine plants will close earlier than planned, in 2022.

One factor that has inhibited efficient employment of new renewable energy has been the lack of an accompanying investment in power infrastructure to bring the power to market. It is believed 8,300 km of power lines must be built or upgraded. The different German States have varying attitudes to the construction of new power lines. Industry has had their rates frozen and so the increased costs of the *Energiewende* have been passed on to consumers, who have had rising electricity bills.

Voluntary Market Mechanisms for Renewable Electricity

Voluntary markets, also referred to as green power markets, are driven by consumer preference. Voluntary markets allow a consumer to choose to do more than policy decisions require and reduce the environmental impact of their electricity use. Voluntary green power products must offer a significant benefit and value to buyers to be successful. Benefits may include zero or reduced greenhouse gas emissions, other pollution reductions or other environmental improvements on power stations.

The driving force behind voluntary green electricity within the EU are the liberalized electricity markets and the RES Directive. According to the directive the EU Member States must ensure that the origin of electricity produced from renewables can be guaranteed and therefore a "guarantee of origin" must be issued (article 15). Environmental organisations are using the voluntary market to create new renewables and improving sustainability of the existing power production. In the US the main tool to track and stimulate voluntary actions is Green-e program managed by Center for Resource Solutions. In Europe the main voluntary tool used by the NGOs to promote sustainable electricity production is EKOenergy label.

Recent Developments

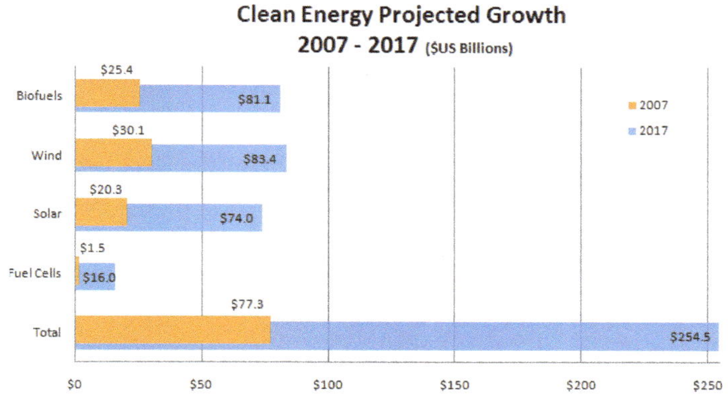

Projected renewable energy investment growth globally (2007–2017)

A number of events in 2006 pushed renewable energy up the political agenda, including the US mid-term elections in November, which confirmed clean energy as a mainstream issue. Also in 2006, the Stern Review made a strong economic case for investing in low carbon technologies now, and argued that economic growth need not be incompatible with cutting energy consumption. According to a trend analysis from the United Nations Environment Programme, climate change concerns coupled with recent high oil prices and increasing government support are driving increasing rates of investment in the renewable energy and energy efficiency industries.

Investment capital flowing into renewable energy reached a record US$77 billion in 2007, with the upward trend continuing in 2008. The OECD still dominates, but there is now increasing activity from companies in China, India and Brazil. Chinese companies were the second largest recipient of venture capital in 2006 after the United States. In the same year, India was the largest net buyer of companies abroad, mainly in the more established European markets.

New government spending, regulation, and policies helped the industry weather the 2009 economic crisis better than many other sectors. Most notably, U.S. President Barack Obama's American Recovery and Reinvestment Act of 2009 included more than $70 billion in direct spending and tax credits for clean energy and associated transportation programs. This policy-stimulus combination represents the largest federal commitment in U.S. history for renewables, advanced transportation, and energy conservation initiatives. Based on these new rules, many more utilities strengthened their clean-energy programs. Clean Edge suggests that the commercialization of clean energy will help countries around the world deal with the current economic malaise. Once-promising solar energy company, Solyndra, became involved in a political controversy involving U.S. President Barack Obama's administration's authorization of a $535 million loan guarantee to the Corporation in 2009 as part of a program to promote alternative energy growth. The company ceased all business activity, filed for Chapter 11 bankruptcy, and laid-off nearly all of its employees in early September 2011.

In his 24 January 2012, State of the Union address, President Barack Obama restated his commitment to renewable energy. Obama said that he "will not walk away from the promise of clean energy." Obama called for a commitment by the Defense Department to purchase 1,000 MW of renewable energy. He also mentioned the long-standing Interior Department commitment to permit 10,000 MW of renewable energy projects on public land in 2012.

As of 2012, renewable energy plays a major role in the energy mix of many countries globally. Renewables are becoming increasingly economic in both developing and developed countries. Prices for renewable energy technologies, primarily wind power and solar power, continued to drop, making renewables competitive with conventional energy sources. Without a level playing field, however, high market penetration of renewables is still dependent on a robust promotional policies. Fossil fuel subsidies, which are far higher than those for renewable energy, remain in place and quickly need to be phased out.

United Nations' Secretary-General Ban Ki-moon has said that "renewable energy has the ability to lift the poorest nations to new levels of prosperity". In October 2011, he "announced the creation of a high-level group to drum up support for energy access, energy efficiency and greater use of renewable energy. The group is to be co-chaired by Kandeh Yumkella, the chair of UN Energy and director general of the UN Industrial Development Organisation, and Charles Holliday, chairman of Bank of America".

Worldwide use of solar power and wind power continued to grow significantly in 2012. Solar electricity consumption increased by 58 percent, to 93 terawatt-hours (TWh). Use of wind power in 2012 increased by 18.1 percent, to 521.3 TWh. Global solar and wind energy installed capacities continued to expand even though new investments in these technologies declined during 2012. Worldwide investment in solar power in 2012 was $140.4 billion, an 11 percent decline from 2011, and wind power investment was down 10.1 percent, to $80.3 billion. But due to lower production costs for both technologies, total installed capacities grew sharply. This investment decline, but growth in installed capacity, may again occur in 2013. Analysts expect the market to triple by 2030. In 2015, investment in renewables exceeded fossils.

100% Renewable Energy

The incentive to use 100% renewable energy for electricity, transport, or even total primary energy supply globally, has been motivated by global warming and other ecological as well as economic concerns. The Intergovernmental Panel on Climate Change has said that there are few fundamental technological limits to integrating a portfolio of renewable energy technologies to meet most of total global energy demand. In reviewing 164 recent scenarios of future renewable energy growth, the report noted that the majority expected renewable sources to supply more than 17% of total energy by 2030, and 27% by 2050; the highest forecast projected 43% supplied by renewables by 2030 and 77% by 2050. Renewable energy use has grown much faster than even advocates anticipated. At the national level, at least 30 nations around the world already have renewable energy contributing more than 20% of energy supply. Also, Professors S. Pacala and Robert H. Socolow have developed a series of "stabilization wedges" that can allow us to maintain our quality of life while avoiding catastrophic climate change, and "renewable energy sources," in aggregate, constitute the largest number of their "wedges."

Mark Z. Jacobson, professor of civil and environmental engineering at Stanford University and director of its Atmosphere and Energy Program says producing all new energy with wind power, solar power, and hydropower by 2030 is feasible and existing energy supply arrangements could be replaced by 2050. Barriers to implementing the renewable energy plan are seen to be "primarily social and political, not technological or economic". Jacobson says that energy costs with a wind, solar, water system should be similar to today's energy costs.

Similarly, in the United States, the independent National Research Council has noted that "sufficient domestic renewable resources exist to allow renewable electricity to play a significant role in future electricity generation and thus help confront issues related to climate change, energy security, and the escalation of energy costs ... Renewable energy is an attractive option because renewable resources available in the United States, taken collectively, can supply significantly greater amounts of electricity than the total current or projected domestic demand." .

The most significant barriers to the widespread implementation of large-scale renewable energy and low carbon energy strategies are primarily political and not technological. According to the 2013 *Post Carbon Pathways* report, which reviewed many international studies, the key roadblocks are: climate change denial, the fossil fuels lobby, political inaction, unsustainable energy consumption, outdated energy infrastructure, and financial constraints.

Energy Efficiency

Moving towards energy sustainability will require changes not only in the way energy is supplied, but in the way it is used, and reducing the amount of energy required to deliver various goods or services is essential. Opportunities for improvement on the demand side of the energy equation are as rich and diverse as those on the supply side, and often offer significant economic benefits.

A sustainable energy economy requires commitments to both renewables and efficiency. Renewable energy and energy efficiency are said to be the "twin pillars" of sustainable energy policy. The American Council for an Energy-Efficient Economy has explained that both resources must be developed in order to stabilize and reduce carbon dioxide emissions:

Efficiency is essential to slowing the energy demand growth so that rising clean energy supplies can make deep cuts in fossil fuel use. If energy use grows too fast, renewable energy development will chase a receding target. Likewise, unless clean energy supplies come online rapidly, slowing demand growth will only begin to reduce total emissions; reducing the carbon content of energy sources is also needed.

The IEA has stated that renewable energy and energy efficiency policies are complementary tools for the development of a sustainable energy future, and should be developed together instead of being developed in isolation.

Energy Storage

Energy storage is the capture of energy produced at one time for use at a later time. A device that stores energy is sometimes called an accumulator. Energy comes in multiple forms including radiation, chemical, gravitational potential, electrical potential, electricity, elevated temperature, latent heat and kinetic. Energy storage involves converting energy from forms that are difficult to store to more conveniently or economically storable forms. Bulk energy storage is dominated by pumped hydro, which accounts for 99% of global energy storage.

The Llyn Stwlan dam of the Ffestiniog Pumped Storage Scheme in Wales. The lower power station has four water turbines which can generate a total of 360 MW of electricity for several hours, an example of artificial energy storage and conversion.

Some technologies provide short-term energy storage, while others can endure for much longer.

A wind-up clock stores potential energy (in this case mechanical, in the spring tension), a rechargeable battery stores readily convertible chemical energy to operate a mobile phone, and a hydroelectric dam stores energy in a reservoir as gravitational potential energy. Fossil fuels such as coal and gasoline store ancient energy derived from sunlight by organisms that later died, became buried and over time were then converted into these fuels. Food (which is made by the same process as fossil fuels) is a form of energy stored in chemical form.

Ice storage tanks store ice frozen by cheaper energy at night to meet peak daytime demand for cooling. The energy isn't stored directly, but the work-product of consuming energy (pumping away heat) is stored, having the equivalent effect on daytime consumption.

History

Prehistory

The energy present at the initial formation of the universe is stored in stars such as the Sun, and is used by humans directly (e.g. through solar heating or sun tanning), or indirectly (e.g. by growing crops, consuming photosynthesized plants or conversion into electricity in solar cells).

As a purposeful activity, energy storage has existed since pre-history, though it was often not explicitly recognized as such. An example of mechanical energy storage is the use of logs or boulders as defensive measures in ancient forts—the logs or boulders were collected at the top of a hill or wall, and the energy thus stored was used to attack invaders who came within range. Certainly, storing dried wood or another source for fire, or preserving edible food or seeds, in dry or cool areas such as in a cave, under rocks or underground, serves as other examples of energy storage.

Recent History

In the twentieth century grid electrical power was largely generated by burning fossil fuel. When less power was required, less fuel was burned. Concerns with air pollution, energy imports and global warming have spawned the growth of renewable energy such as solar and wind power. Wind power is uncontrolled and may be generating at a time when no additional power is needed. Solar power varies with cloud cover and at best is only available during daylight hours, while demand often peaks after sunset. Interest in storing power from these intermittent sources grows as the renewable energy industry begins to generate a larger fraction of overall energy consumption.

Off grid electrical use was a niche market in the twentieth century, but in the twenty first century it has expanded. Portable devices are in use all over the world. Solar panels are now a common sight in the rural settings worldwide. Access to electricity is now a question of economics, not location. Powering transportation without burning fuel, however, remains in development.

Methods

Outline

The following list includes natural and other non-commercial types of energy storage. in addition to those designed for use in industry and commerce:

- Mechanical
  - Compressed air energy storage (CAES)
  - Fireless locomotive
  - Flywheel energy storage
  - Gravitational potential energy (device)
  - Hydraulic accumulator
  - Liquid nitrogen
  - Pumped-storage hydroelectricity
- Electrical
  - Capacitor
  - Superconducting magnetic energy storage (SMES)
- Biological
  - Glycogen
  - Starch
- Electrochemical
  - Flow battery
  - Rechargeable battery
  - Supercapacitor
  - UltraBattery
- Thermal
  - Brick storage heater
  - Cryogenic liquid air or nitrogen

  - Eutectic system
  - Ice storage air conditioning
  - Molten salt
  - Phase Change Material
  - Seasonal thermal energy storage
  - Solar pond
  - Steam accumulator
  - Thermal energy storage (general)
- Chemical
  - Biofuels
  - Hydrated salts
  - Hydrogen
  - Hydrogen peroxide
  - Power to gas
  - Vanadium pentoxide

Mechanical Storage

Energy can be stored in water pumped to a higher elevation using pumped storage methods and also by moving solid matter to higher locations. Other commercial mechanical methods include compressing air and flywheels that convert electric energy into kinetic energy and then back again when electrical demand peaks.

Hydroelectricity

Hydroelectric dams with reservoirs can be operated to provide peak generation at times of peak demand. Water is stored in the reservoir during periods of low demand and released when demand is high. The net effect is similar to pumped storage, but without the pumping loss.

While a hydroelectric dam does not directly store energy from other generating units, it behaves equivalently by lowering output in periods of excess electricity from other sources. In this mode, dams are one of the most efficient forms of energy storage, because only the timing of its generation changes. Hydroelectric turbines have a start-up time on the order of a few minutes.

Pumped-storage

The Sir Adam Beck Generating Complex at Niagara Falls, Canada, which includes a large pumped storage hydroelectricity reservoir to provide an extra 174 MW of electricity during periods of peak demand.

Worldwide, pumped-storage hydroelectricity (PSH) is the largest-capacity form of active grid energy storage available, and, as of March 2012, the Electric Power Research Institute (EPRI) reports that PSH accounts for more than 99% of bulk storage capacity worldwide, representing around 127,000 MW. PSH reported energy efficiency varies in practice between 70% and 80%, with claims of up to 87%.

At times of low electrical demand, excess generation capacity is used to pump water from a lower source into a higher reservoir. When demand grows, water is released back into a lower reservoir (or waterway or body of water) through a turbine, generating electricity. Reversible turbine-generator assemblies act as both a pump and turbine (usually a Francis turbine design). Nearly all facilities use the height difference between two water bodies. Pure pumped-storage plants shift the water between reservoirs, while the "pump-back" approach is a combination of pumped storage and conventional hydroelectric plants that use natural stream-flow.

Compressed Air

A compressed air locomotive used inside a mine between 1928 and 1961.

Compressed air energy storage (CAES) uses surplus energy to compress air for subsequent electricity generation. Small scale systems have long been used in such applications as propulsion of mine locomotives. The compressed air is stored in an underground reservoir.

Compression of air creates heat; the air is warmer after compression. Expansion requires heat. If no extra heat is added, the air will be much colder after expansion. If the heat generated during compression can be stored and used during expansion, efficiency improves considerably. A CAES system can deal with the heat in three ways. Air storage can be adiabatic, diabatic, or isothermal. Another approach uses compressed air to power vehicles.

Flywheel Energy Storage

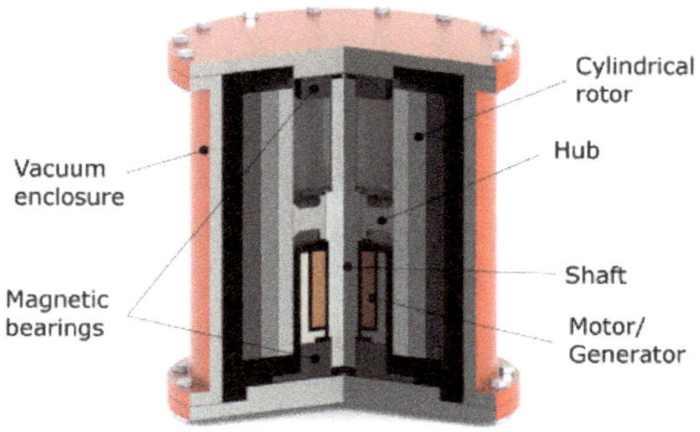

The main components of a typical flywheel.

Flywheel energy storage (FES) works by accelerating a rotor (flywheel) to a very high speed, holding energy as rotational energy. When energy is extracted, the flywheel's rotational speed declines as a consequence of conservation of energy; adding energy correspondingly results in an increase in the speed of the flywheel.

Most FES systems use electricity to accelerate and decelerate the flywheel, but devices that directly use mechanical energy are under consideration.

FES systems have rotors made of high strength carbon-fiber composites, suspended by magnetic bearings and spinning at speeds from 20,000 to over 50,000 rpm in a vacuum enclosure. Such flywheels can reach maximum speed ("charge") in a matter of minutes. The flywheel system is connected to a combination electric motor/generator.

FES systems have relatively long lifetimes (lasting decades with little or no maintenance; full-cycle lifetimes quoted for flywheels range from in excess of 10^5, up to 10^7, cycles of use), high energy density (100–130 W·h/kg, or 360–500 kJ/kg) and power density.

A Flybrid Kinetic Energy Recovery System flywheel. Built for use on Formula 1 racing cars, it is employed to recover and reuse kinetic energy captured during braking.

Gravitational Potential Energy Storage With Solid Masses

Changing the altitude of solid masses can store or release energy via an elevating system driven by an electric motor/generator.

Companies such as Energy Cache and Advanced Rail Energy Storage (ARES) are working on this. ARES uses rails to move concrete weights up and down. Stratosolar proposes to use winches supported by buoyant platforms at an altitude of 20 kilometers, to raise and lower solid masses. Sink Float Solutions proposes to use winches supported by an ocean barge for taking advantage of a 4 km (13,000 ft) elevation difference between the surface and the seabed. ARES estimated a capital cost for the storage capacity of around 60% of pump storage hydroelectricity, Stratosolar $100/kWh and Sink Float Solutions $25/kWh (4000 m depth) and $50/kWh (with 2000 m depth).

Potential energy storage or gravity energy storage was under active development in 2013 in association with the California Independent System Operator. It examined the movement of earth-filled hopper rail cars driven by electric locomotives) from lower to higher elevations.

ARES claimed advantages including indefinite storage with no energy losses, low costs when earth/rocks are used and conservation of water resources.

Thermal Storage

Thermal storage is the temporary storage or removal of heat. TES is practical because of water's large heat of fusion: the melting of one metric ton of ice (approximately one cubic metre in size) can capture 334 megajoules [MJ] (317,000 BTU) of thermal energy.

An example is Alberta, Canada's Drake Landing Solar Community, for which 97% of the year-round heat is provided by solar-thermal collectors on the garage roofs, with a borehole thermal energy store (BTES) being the enabling technology. STES projects often have paybacks in the four-to-six year range. In Braestrup, Denmark, the community's solar district heating system also utilizes STES, at a storage temperature of 65 °C (149 °F). A heat pump, which is run only when there is surplus wind power available on the national grid, is used to raise the temperature to 80 °C (176 °F) for distribution. When surplus wind generated electricity is not available, a gas-fired boiler is used. Twenty percent of Braestrup's heat is solar.

District heating accumulation tower from Theiss near Krems an der Donau in Lower Austria with a thermal capacity of 2 GWh

Latent Heat Thermal Energy Storage (LHTES)

Latent heat thermal energy storage systems works with materials with high latent heat (heat of fusion) capacity, known as phase change materials (PCMs). The main advantage of these materials is that their latent heat storage capacity is much more than sensible heat. In a specific temperature range, phase changes from solid to liquid absorbs a large amount of thermal energy for later use.

Electrochemical

Rechargeable Battery

A rechargeable battery bank used as an uninterruptible power supply in a data center

A rechargeable battery, comprises one or more electrochemical cells. It is known as a 'secondary cell' because its electrochemical reactions are electrically reversible. Rechargeable batteries come in many different shapes and sizes, ranging from button cells to megawatt grid systems.

Rechargeable batteries have lower total cost of use and environmental impact than non-rechargeable (disposable) batteries. Some rechargeable battery types are available in the same form factors as disposables. Rechargeable batteries have higher initial cost but can be recharged very cheaply and used many times.

Common rechargeable battery chemistries include:

- Lead–acid battery: Lead acid batteries hold the largest market share of electric storage products. A single cell produces about 2V when charged. In the charged state the metallic lead negative electrode and the lead sulfate positive electrode are immersed in a dilute sulfuric acid (H2SO4) electrolyte. In the discharge process electrons are pushed out of the cell as lead sulfate is formed at the negative electrode while the electrolyte is reduced to water.
- Nickel–cadmium battery (NiCd): Uses nickel oxide hydroxide and metallic cadmium as electrodes. Cadmium is a toxic element, and was banned for most uses by the European Union in 2004. Nickel–cadmium batteries have been almost completely replaced by nickel–metal hydride (NiMH) batteries.
- Nickel–metal hydride battery (NiMH): First commercial types were available in 1989. These are now a common consumer and industrial type. The battery has a hydrogen-absorbing alloy for the negative electrode instead of cadmium.
- Lithium-ion battery: The choice in many consumer electronics and have one of the best energy-to-mass ratios and a very slow self-discharge when not in use.
- Lithium-ion polymer battery: These batteries are light in weight and can be made in any shape desired.

Flow Battery

A flow battery operates by passing a solution over a membrane where ions are exchanged to charge/discharge the cell. Cell voltage is chemically determined by the Nernst equation and ranges, in practical applications, from 1.0 to 2.2 V. Its storage capacity is a function of the volume of the tanks holding the solution.

A flow battery is technically akin both to a fuel cell and an electrochemical accumulator cell. Commercial applications are for long half-cycle storage such as backup grid power.

Supercapacitor

Supercapacitors, also called electric double-layer capacitors (EDLC) or ultracapacitors, are generic terms for a family of electrochemical capacitors that do not have conventional solid dielectrics. Capacitance is determined by two storage principles, double-layer capacitance and pseudocapacitance.

One of a fleet of electric capabuses powered by supercapacitors, at a quick-charge station-bus stop, in service during Expo 2010 Shanghai China. Charging rails can be seen suspended over the bus.

Supercapacitors bridge the gap between conventional capacitors and rechargeable batteries. They store the most energy per unit volume or mass (energy density) among capacitors. They support up to 10,000 farads/1.2 volt, up to 10,000 times that of electrolytic capacitors, but deliver or accept less than half as much power per unit time (power density).

While supercapacitors have energy densities that are approximately 10% of batteries, their power density is generally 10 to 100 times greater. This results in much shorter charge/discharge cycles. Additionally, they will tolerate many more charge and discharge cycles than batteries.

Supercapacitors support a broad spectrum of applications, including:

- Low supply current for memory backup in static random-access memory (SRAM)
- Power for cars, buses, trains, cranes and elevators, including energy recovery from braking, short-term energy storage and burst-mode power delivery

UltraBattery

The UltraBattery is a hybrid lead-acid cell and carbon-based ultracapacitor (or supercapacitor) invented by Australia's national research body, the Commonwealth Scientific and Industrial Research Organisation (CSIRO). The lead-acid cell and ultracapacitor share the sulfuric acid electrolyte and both are packaged into the same physical unit. The UltraBattery can be manufactured with similar physical and electrical characteristics to conventional lead-acid batteries making it possible to cost-effectively replace many lead-acid applications.

The UltraBattery tolerates high charge and discharge levels and endures large numbers of cycles, outperforming previous lead-acid cells by more than an order of magnitude. In hybrid-electric vehicle tests, millions of cycles have been achieved. The UltraBattery is also highly tolerant to the effects of sulfation compared with traditional lead-acid cells. This means it can operate continuously in partial state of charge whereas traditional lead-acid batteries are generally held at full charge between discharge events. It is generally electrically inefficient to fully charge a lead-acid battery so by decreasing time spent in the top region of charge the UltraBattery achieves high efficiencies, typically between 85 and 95% DC-DC.

The UltraBattery can work across a wide range of applications. The constant cycling and fast charging and discharging necessary for applications such as grid regulation and leveling and electric vehicles can damage chemical batteries, but are well handled by the ultracapacitive qualities of UltraBattery technology. The technology has been installed in Australia and the US on the megawatt scale, performing frequency regulation and renewable smoothing applications.

Other Chemical

Power to gas

Power to gas is a technology which converts electricity into a gaseous fuel such as hydrogen or methane. The three commercial methods use electricity to reduce water into hydrogen and oxygen by means of electrolysis.

In the first method, hydrogen is injected into the natural gas grid or is used in transport or industry. The second method is to combine the hydrogen with carbon dioxide to produce methane using a methanation reaction such as the Sabatier reaction, or biological methanation, resulting in an extra energy conversion loss of 8%. The methane may then be fed into the natural gas grid. The third method uses the output gas of a wood gas generator or a biogas plant, after the biogas upgrader is mixed with the hydrogen from the electrolyzer, to upgrade the quality of the biogas.

Hydrogen

The element hydrogen can be a form of stored energy. Hydrogen can produce electricity via a hydrogen fuel cell.

At penetrations below 20% of the grid demand, renewables do not severely change the economics; but beyond about 20% of the total demand, external storage becomes important. If these sources are used to make ionic hydrogen, they can be freely expanded. A 5-year community-based pilot program using wind turbines and hydrogen generators began in 2007 in the remote community of Ramea, Newfoundland and Labrador. A similar project began in 2004 on Utsira, a small Norwegian island.

Energy losses involved in the hydrogen storage cycle come from the electrolysis of water, liquification or compression of the hydrogen and conversion to electricity.

About 50 kW·h (180 MJ) of solar energy is required to produce a kilogram of hydrogen, so the cost of the electricity is crucial. At $0.03/kWh, a common off-peak high-voltage line rate in the United States, hydrogen costs $1.50 a kilogram for the electricity, equivalent to $1.50/gallon for gasoline. Other costs include the electrolyzer plant, hydrogen compressors or liquefaction, storage and transportation.

Underground hydrogen storage is the practice of hydrogen storage in underground caverns, salt domes and depleted oil and gas fields. Large quantities of gaseous hydrogen have been stored in underground caverns by Imperial Chemical Industries for many years without any difficulties. The European Hyunder project indicated in 2013 that storage of wind and solar energy using underground hydrogen would require 85 caverns.

Methane

Methane is the simplest hydrocarbon with the molecular formula CH_4. Methane is more easily stored than hydrogen and the transportation. Storage and combustion infrastructure (pipelines, gasometers, power plants) are mature.

Synthetic natural gas (syngas or SNG) can be created in a multi-step process, starting with hydrogen and oxygen. Hydrogen is then reacted with carbon dioxide in a Sabatier process, producing methane and water. Methane can be stored and later used to produce electricity. The resulting water is recycled, reducing the need for water. In the electrolysis stage oxygen is stored for methane combustion in a pure oxygen environment at an adjacent power plant, eliminating nitrogen oxides.

Methane combustion produces carbon dioxide (CO_2) and water. The carbon dioxide can be recycled to boost the Sabatier process and water can be recycled for further electrolysis. Methane production, storage and combustion recycles the reaction products.

The CO_2 has economic value as a component of an energy storage vector, not a cost as in carbon capture and storage.

Power to Liquid

Power to liquid is similar to power to gas, however the hydrogen produced by electrolysis from wind and solar electricity isn't converted into gases such as methane but into liquids such as methanol. Methanol is easier in handling than gases and requires less safety precautions than hydrogen. It can be used for transportation, including aircraft, but also for industrial purposes or in the power sector.

Biofuels

Various biofuels such as biodiesel, vegetable oil, alcohol fuels, or biomass can replace fossil fuels. Various chemical processes can convert the carbon and hydrogen in coal, natural gas, plant and animal biomass and organic wastes into short hydrocarbons suitable as replacements for existing hydrocarbon fuels. Examples are Fischer–Tropsch diesel, methanol, dimethyl ether and syngas. This diesel source was used extensively in World War II in Germany, which faced limited access to crude oil supplies. South Africa produces most of the country's diesel from coal for similar reasons. A long term oil price above US$35/bbl may make such large scale synthetic liquid fuels economical.

Aluminium, Boron, Silicon, and Zinc

Aluminium, Boron, silicon, lithium, and zinc have been proposed as energy storage solutions.

Electrical Methods

Capacitor

This mylar-film, oil-filled capacitor has very low inductance and low resistance, to provide the high-power (70 megawatts) and the very high speed (1.2 microsecond) discharges needed to operate a dye laser.

A capacitor (originally known as a 'condenser') is a passive two-terminal electrical component used to store energy electrostatically. Practical capacitors vary widely, but all contain at least two electrical conductors (plates) separated by a dielectric (i.e., insulator). A capacitor can store electric energy when disconnected from its charging circuit, so it can be used like a temporary battery, or like other types of rechargeable energy storage system. Capacitors are commonly used in electronic devices to maintain power supply while batteries change. (This prevents loss of information in volatile memory.)

Conventional capacitors provide less than 360 joules per kilogram, while a conventional alkaline battery has a density of 590 kJ/kg.

Capacitors store energy in an electrostatic field between their plates. Given a potential difference across the conductors (e.g., when a capacitor is attached across a battery), an electric field develops across the dielectric, causing positive charge (+Q) to collect on one plate and negative charge (-Q) to collect on the other plate. If a battery is attached to a capacitor for a sufficient amount of time, no current can flow through the capacitor. However, if an accelerating or alternating voltage is applied across the leads of the capacitor, a displacement current can flow.

Capacitance is greater given a narrower separation between conductors and when the conductors have a larger surface area. In practice, the dielectric between the plates emits a small amount of leakage current and has an electric field strength limit, known as the breakdown voltage. The conductors and leads introduce undesired inductance and resistance.

Research is assessing the quantum effects of nanoscale capacitors for digital quantum batteries.

Superconducting Magnetics

Superconducting magnetic energy storage (SMES) systems store energy in a magnetic field created by the flow of direct current in a superconducting coil that has been cooled to a temperature below its superconducting critical temperature. A typical SMES system includes a superconducting coil, power conditioning system and refrigerator. Once the superconducting coil is charged, the current does not decay and the magnetic energy can be stored indefinitely.

The stored energy can be released to the network by discharging the coil. The associated inverter/rectifier accounts for about 2–3% energy loss in each direction. SMES loses the least amount of electricity in the energy storage process compared to other methods of storing energy. SMES systems offer round-trip efficiency greater than 95%.

Due to the energy requirements of refrigeration and the cost of superconducting wire, SMES is used for short duration storage such as improving power quality. It also has applications in grid balancing.

Interseasonal Thermal Storage

Seasonal thermal energy storage (STES) allows heat or cold to be used months after it was collected from waste energy or natural sources. The material can be stored in contained aquifers, clusters of boreholes in geological substrates such as sand or crystalline bedrock, in lined pits filled with gravel and water, or water-filled mines.

Applications

Mills

A more recent application is the control of waterways to drive water mills for processing grain or powering machinery. Complex systems of reservoirs and dams were constructed to store and release water (and the potential energy it contained) when required.

Home Energy Storage

Home energy storage is expected to become increasingly present given the growing importance of distributed generation (especially photovoltaics) and the important share of energy consumption in buildings. A household equipped with photovoltaics can achieve a maximum electricity self-sufficiency of about 40%. To reach higher levels of self-sufficiency,energy storage is needed, given the mismatch between energy consumption and energy production from photovoltaics. In 2015, multiple manufacturers announced rechargeable battery systems for storing energy, generally to hold surplus energy from home solar/wind generation.

Tesla Motors announced the first two models of the Tesla Powerwall. One is a 10 kWh weekly cycle version for backup applications and the other is a 7 kWh version for daily cycle applications.

Enphase Energy announced an integrated system that allows home users to store, monitor and manage electricity. The system stores 1.2 kWh hours of energy and 275W/500W power output.

Grid Electricity

Renewable Energy Storage

Construction of the Salt Tanks which provide efficient thermal energy storage so that output can be provided after the sun goes down, and output can be scheduled to meet demand requirements. The 280 MW Solana Generating Station is designed to provide six hours of energy storage. This allows the plant to generate about 38 percent of its rated capacity over the course of a year.

The largest source and the greatest store of renewable energy is provided by hydroelectric dams. A large reservoir behind a dam can store enough water to average the annual flow of a river between dry and wet seasons. A very large reservoir can store enough water to average the flow of a river between dry and wet years. While a hydroelectric dam does not directly store energy from intermittent sources, it does balance the grid by lowering its output and retaining its water when power is generated by solar or wind. If wind or solar generation exceeds the regions hydroelectric capacity, then some additional form of energy balancing will be needed.

Many renewable energy sources (notably solar and wind) produce variable power. Storage systems can level out the imbalances between supply and demand that this causes. Electricity must be used as it is generated or converted immediately into storable forms.

The first method of large scale storage was pumped-storage hydroelectricity. Areas of the world such as Norway, Washington and Oregon states and Wales have used geographic features to supply elevated reservoirs, using electricity to fill them. When needed, the water passes through generators and converts the potential energy back to electricity. Pumped storage in Norway, which gets almost all its electricity from hydro, has an instantaneous capacity of 25–30 GW expandable to 60 GW—enough to be "Europe's battery".

Common forms of renewable energy storage include pumped-storage hydroelectric dams, rechargeable batteries, thermal storage including molten salts which can efficiently store and release very large quantities of heat energy, and compressed air energy storage, flywheels, cryogenic systems and superconducting magnetic coils.

Surplus power can also be converted into methane (sabatier process) with stockage in the natural gas network.

In 2011, the Bonneville Power Administration in Northwestern United States created an experimental program to absorb excess wind and hydro power generated at night or during stormy periods that are accompanied by high winds. Under central control, home appliances absorb surplus energy by heating ceramic bricks in special space heaters to hundreds of degrees and by boosting the temperature of modified hot water heater tanks. After charging, the appliances provide home heating and hot water as needed. The experimental system was created as a result of a severe 2010 storm that overproduced renewable energy to the extent that all conventional power sources were shut down, or in the case of a nuclear power plant, reduced to its lowest possible operating level, leaving a large area running almost completely on renewable energy.

Another advanced method used at the Solar Project in the United States and the Solar Tres Power Tower in Spain uses molten salt to store thermal energy captured from the sun and then convert it and dispatch it as electrical power. The system pumps molten salt through a tower or other special conduits to be heated by the sun. Insulated tanks store the solution. Electricity is produced by turning water to steam that is fed to turbines.

Since the early 21st century batteries have been applied to utility scale load-leveling and frequency regulation capabilities.

In vehicle-to-grid storage, electric vehicles that are plugged into the energy grid can deliver stored electrical energy from their batteries into the grid when needed.

Generation

Chemical fuels remain the dominant form of energy storage for electricity generation. Natural gas is crowding out other forms, such as oil and coal.

Air Conditioning

thermal energy storage (TES) can be used for air conditioning. It is most widely used for cooling single large buildings and/or groups of smaller buildings. Commercial air conditioning systems are the biggest contributors to peak electrical loads. In 2009, thermal storage was used in over 3,300 buildings in over 35 countries. It works by creating ice at night and using the ice to for cooling during the hotter daytime periods.

The most popular technique is ice storage, which requires less space than water and is less costly than fuel cells or flywheels. In this application, a standard chiller runs at night to produce an ice pile. Water then circulates through the pile during the day to chill water that would normally be the chiller's daytime output.

A partial storage system minimizes capital investment by running the chillers nearly 24 hours a day. At night, they produce ice for storage and during the day they chill water. Water circulating through the melting ice augments the production of chilled water. Such a system makes ice for 16 to 18 hours a day and melts ice for six hours a day. Capital expenditures are reduced because the chillers can be just 40 - 50% of the size needed for a conventional, no-storage design. Storage sufficient to store half a day's available heat is usually adequate.

A full storage system shuts off the chillers during peak load hours. Capital costs are higher, as such a system requires larger chillers and a larger ice storage system.

This ice is produced when electrical utility rates are lower. Off-peak cooling systems can lower energy costs. The U.S. Green Building Council has developed the Leadership in Energy and Environmental Design (LEED) program to encourage the design of reduced-environmental impact buildings. Off-peak cooling may help toward LEED Certification.

Thermal storage for heating is less common than for cooling. An example of thermal storage is storing solar heat to be used for heating at night.

Latent heat can also be stored in technical phase change materials (PCMs). These can be encapsulated in wall and ceiling panels, to moderate room temperatures.

Transport

Liquid hydrocarbon fuels are the most commonly used forms of energy storage for use in transportation. Other energy carriers such as hydrogen can be used to avoid producing greenhouse gases.

Electronics

Capacitors are widely used in electronic circuits for blocking direct current while allowing alternating current to pass. In analog filter networks, they smooth the output of power supplies. In resonant circuits they tune radios to particular frequencies. In electric power transmission systems they stabilize voltage and power flow.

Use Cases

The United States Department of Energy International Energy Storage Database (IESDB), is a free-access database of energy storage projects and policies funded by the United States Department of Energy Office of Electricity and Sandia National Labs.

Economics

The economics of Energy Storage strictly depends on the reserve service requested, and several uncertainty factors affect the profitability of Energy Storage. Therefore not every Energy Storage is technically and economically suitable for the storage of several MWh, and the optimal size of the Energy Storage is market and location dependent.

Moreover, ESS are affected by several risks, e.g.:

1) techno-economic risks, which are related to the specific technology;

2) Market risks, which are the factors that affect the electricity supply system;

3) Regulation and policy risks.

Therefore, traditional techniques based on deterministic Discounted Cash Flow (DCF) for the investment appraisal are not fully adequate to evaluate these risks and uncertainties and the investor's flexibility to deal with them. Hence, the literature recommends to assess the value of risks and uncertainties through the Real Option Analysis (ROA), which is a valuable method in uncertain contexts.

The economic valuation of large-scale applications (including pumped hydro storage and compressed air) considers benefits including: wind curtailment avoidance, grid congestion avoidance, price arbitrage and carbon free energy delivery. In one technical assessment by the Carnegie Mellon Electricity Industry Centre, economic goals could be met with batteries if energy storage were achievable at a capital cost of $30 to $50 per kilowatt-hour of storage capacity.

A metric for calculating the energy efficiency of storage systems is Energy Storage On Energy Invested (ESOI) which is the useful energy used to make the storage system divided into the lifetime energy storage. For lithium ion batteries this is around 10, and for lead acid batteries it is about 2. Other forms of storage such as pumped hydroelectric storage generally have higher ESOI, such as 210.

Research

Germany

The German Federal government has allocated €200M (approximately US$270M) for advanced research, as well as providing a further €50M to subsidize battery storage for use with residential rooftop solar panels, according to a representative of the German Energy Storage Association.

Siemens AG commissioned a production-research plant to open in 2015 at the *Zentrum für Sonnenenergie und Wasserstoff (ZSW,* the German Center for Solar Energy and Hydrogen Research in the State of Baden-Württemberg), a university/industry collaboration in Stuttgart, Ulm and Widderstall, staffed by approximately 350 scientists, researchers, engineers, and technicians. The plant develops new near-production manufacturing materials and processes (NPMM&P) using a computerized Supervisory Control and Data Acquisition (SCADA) system. Its goals will enable the expansion of rechargeable battery production with both increased quality and reduced manufacturing costs.

United States

In 2014, research and test centers opened to evaluate energy storage technologies. Among them was the Advanced Systems Test Laboratory at the University of Wisconsin at Madison in Wisconsin State, which partnered with battery manufacturer Johnson Controls. The laboratory was created as part of the university's newly opened Wisconsin Energy Institute. Their goals include the evaluation of state-of-the-art and next generation electric vehicle batteries, including their use as grid supplements.

The State of New York unveiled its New York Battery and Energy Storage Technology (NY-BEST) Test and Commercialization Center at Eastman Business Park in Rochester, New York, at a cost of $23 million for its almost 1,700 m² laboratory. The center includes the Center for Future Energy Systems, a collaboration between Cornell University of Ithaca, New York and the Rensselaer Polytechnic Institute in Troy, New York. NY-BEST tests, validates and independently certifies diverse forms of energy storage intended for commercial use.

United Kingdom

In the United Kingdom, some fourteen industry and government agencies allied with seven British universities in May 2014 to create the SUPERGEN Energy Storage Hub in order to assist in the coordination of energy storage technology research and development.

References

- Lynn R. Kahle, Eda Gurel-Atay, Eds (2014). Communicating Sustainability for the Green Economy. New York: M.E. Sharpe. ISBN 978-0-7656-3680-5.
- Huesemann, Michael H., and Joyce A. Huesemann (2011). Technofix: Why Technology Won't Save Us or the Environment, Chapter 5, "In Search of Solutions: Efficiency Improvements", New Society Publishers, ISBN 978-0-86571-704-6.
- Taylor, John W R (1974). Jane's All the World's Aircraft 1974-75. London: Jane's Yearbooks. p. 573. ISBN 0 354 00502 2.
- Spellman, Frank R. (2013). Safe Work Practices for Green Energy Jobs (first ed.). DEStech Publications. p. 323. ISBN 978-1-60595-075-4. Retrieved 29 December 2014.
- REN21 (2014). "Renewables 2014 Global Status Report" (PDF). (Paris: REN21 Secretariat). ISBN 978-3-9815934-2-6.
- Buchan, David (June 2012). The Energiewende — Germany's gamble (PDF). Oxford, UK: Oxford Institute for Energy Studies. ISBN 978-1-907555-52-7. Retrieved 2016-05-12.
- B. E. Conway (1999). Electrochemical Supercapacitors: Scientific Fundamentals and Technological Applications. Berlin: Springer. ISBN 0306457369. Retrieved May 2, 2013.
- Aifantis, Katerina E.; Hackney, Stephen A.; Kumar, R. Vasant (March 30, 2010). High Energy Density Lithium Batteries: Materials, Engineering, Applications. John Wiley & Sons. ISBN 978-3-527-63002-8.
- Bird, John (2010). Electrical and Electronic Principles and Technology. Routledge. pp. 63–76. ISBN 9780080890562. Retrieved March 17, 2013.
- "Northrop Grumman Corporation: News and Events - News Release". investor.northropgrumman.com. Retrieved 2016-05-22.

3

Energy Efficiency: An Overview

Efficient energy use is the aim of reducing the usage of energy and provides products that help the same, for example, the usage of natural skylights. The section focuses on other aspects of efficient use such as energy efficiency in transportation, energy efficiency implementation, energy intensity and energy efficiency gap.

Efficient Energy Use

Efficient energy use, sometimes simply called energy efficiency, is the goal to reduce the amount of energy required to provide products and services. For example, insulating a home allows a building to use less heating and cooling energy to achieve and maintain a comfortable temperature. Installing fluorescent lights, LED lights or natural skylights reduces the amount of energy required to attain the same level of illumination compared with using traditional incandescent light bulbs. Improvements in energy efficiency are generally achieved by adopting a more efficient technology or production process or by application of commonly accepted methods to reduce energy losses.

A spiral-type integrated compact fluorescent lamp, which has been in popular use among North American consumers since its introduction in the mid-1990s.

There are many motivations to improve energy efficiency. Reducing energy use reduces energy costs and may result in a financial cost saving to consumers if the energy savings offset any additional costs of implementing an energy efficient technology. Reducing energy use is also seen as a solution to the problem of reducing greenhouse gas emissions. According to the International Energy Agency, improved energy efficiency in buildings, industrial processes and transportation could reduce the world's energy needs in 2050 by one third, and help control global emissions of greenhouse gases.

Energy efficiency and renewable energy are said to be the *twin pillars* of sustainable energy policy and are high priorities in the sustainable energy hierarchy. In many countries energy efficiency is also seen to have a national security benefit because it can be used to reduce the level of energy imports from foreign countries and may slow down the rate at which domestic energy resources are depleted.

Overview

Energy efficiency has proved to be a cost-effective strategy for building economies without necessarily increasing energy consumption. For example, the state of California began implementing energy-efficiency measures in the mid-1970s, including building code and appliance standards with strict efficiency requirements. During the following years, California's energy consumption has remained approximately flat on a per capita basis while national US consumption doubled. As part of its strategy, California implemented a "loading order" for new energy resources that puts energy efficiency first, renewable electricity supplies second, and new fossil-fired power plants last. States such as Connecticut and New York have created quasi-public Green Banks to help residential and commercial building-owners finance energy efficiency upgrades that reduce emissions and cut consumers' energy costs.

Lovin's Rocky Mountain Institute points out that in industrial settings, "there are abundant opportunities to save 70% to 90% of the energy and cost for lighting, fan, and pump systems; 50% for electric motors; and 60% in areas such as heating, cooling, office equipment, and appliances." In general, up to 75% of the electricity used in the US today could be saved with efficiency measures that cost less than the electricity itself. The same holds true for this is home and there is 78% of electricity uses D in your home-owners, leaky ducts have remained an invisible energy culprit for years. In fact, researchers at the US Department of Energy and their consortium, Residential Energy Efficient Distribution Systems (REEDS) have found that duct efficiency may be as low as 50–70%. The US Department of Energy has stated that there is potential for energy saving in the magnitude of 90 Billion kWh by increasing home energy efficiency.

Other studies have emphasized this. A report published in 2006 by the McKinsey Global Institute, asserted that "there are sufficient economically viable opportunities for energy-productivity improvements that could keep global energy-demand growth at less than 1 percent per annum"—less than half of the 2.2 percent average growth anticipated through 2020 in a business-as-usual scenario. Energy productivity, which measures the output and quality of goods and services per unit of energy input, can come from either reducing the amount of energy required to produce something, or from increasing the quantity or quality of goods and services from the same amount of energy.

The Vienna Climate Change Talks 2007 Report, under the auspices of the United Nations Framework Convention on Climate Change (UNFCCC), clearly shows "that energy efficiency can achieve real emission reductions at low cost."

Appliances

Modern appliances, such as, freezers, ovens, stoves, dishwashers, and clothes washers and dryers, use significantly less energy than older appliances. Installing a clothesline will significantly reduce one's energy consumption as their dryer will be used less. Current energy efficient refrigerators, for example, use 40 percent less energy than conventional models did in 2001. Following this, if all households in Europe changed their more than ten-year-old appliances into new ones, 20 billion kWh of electricity would be saved annually, hence reducing CO_2 emissions by almost 18 billion kg. In the US, the corresponding figures would be 17 billion kWh of electricity and 27,000,000,000 lb (1.2×10^{10} kg) CO_2. According to a 2009 study from McKinsey & Company the replacement of old appliances is one of the most efficient global measures to reduce emissions of greenhouse gases. Modern power management systems also reduce energy usage by idle appliances by turning them off or putting them into a low-energy mode after a certain time. Many countries identify energy-efficient appliances using energy input labeling.

The impact of energy efficiency on peak demand depends on when the appliance is used. For example, an air conditioner uses more energy during the afternoon when it is hot. Therefore, an energy efficient air conditioner will have a larger impact on peak demand than off-peak demand. An energy efficient dishwasher, on the other hand, uses more energy during the late evening when people do their dishes. This appliance may have little to no impact on peak demand.

Building Design

Receiving a Gold rating for energy and environmental design in September 2011, the Empire State Building is the tallest and largest LEED certified building in the United States and Western Hemisphere., though it will likely be overtaken by New York's own One World Trade Center.

Buildings are an important field for energy efficiency improvements around the world because of their role as a major energy consumer. However, the question of energy use in buildings is not straightforward as the indoor conditions that can be achieved with energy use vary a lot. The measures that keep buildings comfortable, lighting, heating, cooling and ventilation, all consume energy. Typically the level of energy efficiency in a building is measured by dividing energy consumed with the floor area of the building resulting in specific energy consumption (SEC):

$$\frac{Energy\ consumed}{Built\ area}$$

However, the issue is more complex as building materials have embodied energy in them. On the other hand, energy can be recovered from the materials when the building is dismantled by reusing materials or burning them for energy. Moreover, the when the building is used, the indoor conditions can vary resulting in higher and lower quality indoor environments. Finally, overall efficiency is affected by the use of the building: is the building occupied most of the time and are spaces efficiently used — or is the building largely empty? It has even been suggested that for a more complete accounting of energy efficiency, SEC should be amended to include these factors:

$$\frac{Embodied\ energy + Energy\ consumed - Energy\ recovered}{Built\ area \times Utilization\ rate \times Quality\ factor}$$

Thus a balanced approach to energy efficiency in buildings should be more comprehensive than simply trying to minimize energy consumed. Issues such as quality of indoor environment and efficiency of space use should be factored in. Thus the measures used to improve energy efficiency can take many different forms. Often they include passive measures that inherently reduce the need to use energy, such as better insulation. Many serve various functions improving the indoor conditions as well as reducing energy use, such as increased use of natural light.

A building's location and surroundings play a key role in regulating its temperature and illumination. For example, trees, landscaping, and hills can provide shade and block wind. In cooler climates, designing northern hemisphere buildings with south facing windows and southern hemisphere buildings with north facing windows increases the amount of sun (ultimately heat energy) entering the building, minimizing energy use, by maximizing passive solar heating. Tight building design, including energy-efficient windows, well-sealed doors, and additional thermal insulation of walls, basement slabs, and foundations can reduce heat loss by 25 to 50 percent.

Dark roofs may become up to 39 °C (70 °F) hotter than the most reflective white surfaces. They transmit some of this additional heat inside the building. US Studies have shown that lightly colored roofs use 40 percent less energy for cooling than buildings with darker roofs. White roof systems save more energy in sunnier climates. Advanced electronic heating and cooling systems can moderate energy consumption and improve the comfort of people in the building.

Proper placement of windows and skylights as well as the use of architectural features that reflect light into a building can reduce the need for artificial lighting. Increased use of natural and task lighting has been shown by one study to increase productivity in schools and offices. Compact fluorescent lights use two-thirds less energy and may last 6 to 10 times longer than incandescent light bulbs. Newer fluorescent lights produce a natural light, and in most applications they are cost effective, despite their higher initial cost, with payback periods as low as a few months.

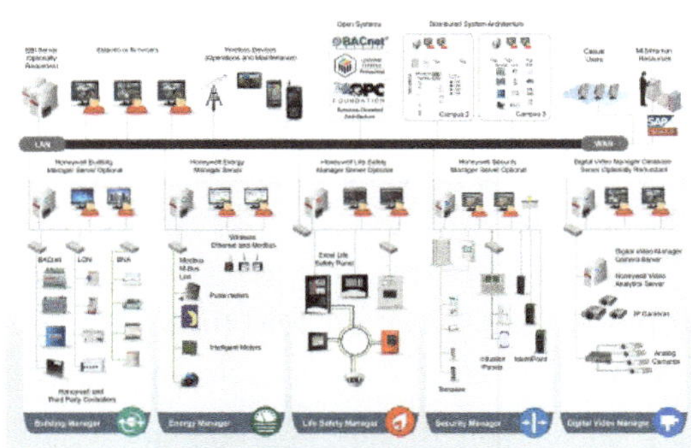

BMS integration

Effective energy-efficient building design can include the use of low cost Passive Infra Reds (PIRs) to switch-off lighting when areas are unnoccupied such as toilets, corridors or even office areas out-of-hours. In addition, lux levels can be monitored using daylight sensors linked to the building's lighting scheme to switch on/off or dim the lighting to pre-defined levels to take into account the natural light and thus reduce consumption. Building Management Systems (BMS) link all of this together in one centralised computer to control the whole building's lighting and power requirements.

In an analysis that integrates a residential bottom-up simulation with an economic multi-sector model, it has been shown that variable heat gains caused by insulation and air-conditioning efficiency can have load-shifting effects that are not uniform on the electricity load. The study also highlighted the impact of higher household efficiency on the power generation capacity choices that are made by the power sector.

The choice of which space heating or cooling technology to use in buildings can have a significant impact on energy use and efficiency. For example, replacing an older 50% efficient natural gas furnace with a new 95% efficient one will dramatically reduce energy use, carbon emissions, and winter natural gas bills. Ground source heat pumps can be even more energy efficient and cost effective. These systems use pumps and compressors to move refrigerant fluid around a thermodynamic cycle in order to "pump" heat against its natural flow from hot to cold, for the purpose of transferring heat into a building from the large thermal reservoir contained within the nearby ground. The end result is that heat pumps typically use four times less electrical energy to deliver an equivalent amount of heat than a direct electrical heater does. Another advantage of a ground source heat pump is that it can be reversed in summertime and operate to cool the air by transferring heat from the building to the ground. The disadvantage of ground source heat pumps is their high initial capital cost, but this is typically recouped within five to ten years as a result of lower energy use.

Smart meters are slowly being adopted by the commercial sector to highlight to staff and for internal monitoring purposes the building's energy usage in a dynamic presentable format. The use of Power Quality Analysers can be introduced into an existing building to assess usage, harmonic distortion, peaks, swells and interruptions amongst others to ultimately make the building more energy-efficient. Often such meters communicate by using wireless sensor networks.

Green Building XML (gbXML) is an emerging schema, a subset of the Building Information Modeling efforts, focused on green building design and operation. gbXML is used as input in several energy simulation engines. But with the development of modern computer technology, a large number of building energy simulation tools are available on the market. When choosing which simulation tool to use in a project, the user must consider the tool's accuracy and reliability, considering the building information they have at hand, which will serve as input for the tool. Yezioro, Dong and Leite developed an artificial intelligence approach towards assessing building performance simulation results and found that more detailed simulation tools have the best simulation performance in terms of heating and cooling electricity consumption within 3% of mean absolute error.

Leadership in Energy and Environmental Design (LEED) is a rating system organized by the US Green Building Council (USGBC) to promote environmental responsibility in building design. They currently offer four levels of certification for existing buildings (LEED-EBOM) and new construction (LEED-NC) based on a building's compliance with the following criteria: Sustainable Sites, Water Efficiency, Energy and Atmosphere, Materials and Resources, Indoor Environmental Quality, and Innovation in Design. In 2013, USGBC developed the LEED Dynamic Plaque, a tool to track building performance against LEED metrics and a potential path to recertification. The following year, the council collaborated with Honeywell to pull data on energy and water use, as well as indoor air quality from a BAS to automatically update the plaque, providing a near-real-time view of performance. The USGBC office in Washington, D.C. is one of the first buildings to feature the live-updating LEED Dynamic Plaque.

A deep energy retrofit is a whole-building analysis and construction process that uses to achieve much larger energy savings than conventional energy retrofits. Deep energy retrofits can be applied to both residential and non-residential ("commercial") buildings. A deep energy retrofit typically results in energy savings of 30 percent or more, perhaps spread over several years, and may significantly improve the building value. The Empire

State Building has undergone a deep energy retrofit process that was completed in 2013. The project team, consisting of representatives from Johnson Controls, Rocky Mountain Institute, Clinton Climate Initiative, and Jones Lang LaSalle will have achieved an annual energy use reduction of 38% and $4.4 million. For example, the 6,500 windows were remanufactured onsite into superwindows which block heat but pass light. Air conditioning operating costs on hot days were reduced and this saved $17 million of the project's capital cost immediately, partly funding other retrofitting. Receiving a gold Leadership in Energy and Environmental Design (LEED) rating in September 2011, the Empire State Building is the tallest LEED certified building in the United States. The Indianapolis City-County Building recently underwent a deep energy retrofit process, which has achieved an annual energy reduction of 46% and $750,000 annual energy saving.

Energy retrofits, including deep, and other types undertaken in residential, commercial or industrial locations are generally supported through various forms of financing or incentives. Incentives include pre-packaged rebates where the buyer/user may not even be aware that the item being used has been rebated or "bought down". "Upstream" or "Midstream" buy downs are common for efficient lighting products. Other rebates are more explicit and transparent to the end user through the use of formal applications. In addition to rebates, which may be offered through government or utility programs, governments sometimes offer tax incentives for energy efficiency projects. Some entities offer rebate and payment guidance and facilitation services that enable energy end use customers tap into rebate and incentive programs.

To evaluate the economic soundness of energy efficiency investments in buildings, cost-effectiveness analysis or CEA can be used. A CEA calculation will produce the value of energy saved, sometimes called *negawatts*, in $/kWh. The energy in such a calculation is virtual in the sense that it was never consumed but rather saved due to some energy efficiency investment being made. Thus CEA allows comparing the price of negawatts with price of energy such as electricity from the grid or the cheapest renewable alternative. The benefit of the CEA approach in energy systems is that it avoids the need to guess future energy prices for the purposes of the calculation, thus removing the major source of uncertainty in the appraisal of energy efficiency investments.

Energy Efficiency By Country

Europe

Energy efficiency targets for 2020 and 2030.

The EU has set itself a 20% energy savings target by 2020 when compared to the projected use of energy in 2020 – roughly equivalent to turning off 400 power stations. At an EU summit in October 2014, EU countries agreed on a new energy efficiency target of 27% or greater by 2030. One mechanism used to achieve the target of 27% is the 'Suppliers Obligations & White Certificates'.

Australia

The Australian national government is actively leading the country in efforts to increase their energy efficiency, mainly through the government's Department of Industry and Science. In July 2009, the Council of Australian Governments, which represents the individual states and territories of Australia, agreed to a National Strategy on Energy Efficiency (NSEE).

This is a ten-year plan accelerate the implementation of a nationwide adoption of energy efficient practices and a preparation for the country's transformation into a low carbon future. There are several different areas of energy use addressed within the NSEE. But, the chapter devoted to the approach on energy efficiency that is to be adopted on a national level stresses four points in achieving stated levels of energy efficiency. They are:

- To help households and businesses transition to a low carbon future
- To streamline the adoption of efficient energy
- To make buildings more energy efficient
- For governments to work in partnership and lead the way to energy efficiency

The overriding agreement that governs this strategy is the National Partnership Agreement on Energy Efficiency.

This document also explains the role of both the commonwealth and the individual states and territories in the NSEE, as well provides for the creation of benchmarks and measurement devices which will transparently show the nation's progress in relation to the stated goals, and addresses the need for funding of the strategy in order to enable it to move forward.

Germany

Energy efficiency is central to energy policy in Germany. As of late 2015, national policy includes the following efficiency and consumption targets (with actual values for 2014):

Efficiency and consumption target	**2014**	**2020**	**2050**
Primary energy consumption (base year 2008)	−8.7%	−20%	−50%
Final energy productivity (2008–2050)	1.6%/year (2008–2014)	2.1%/year (2008–2050)	
Gross electricity consumption (base year 2008)	−4.6%	−10%	−25%
Primary energy consumption in buildings (base year 2008)	−14.8%		−80%
Heat consumption in buildings (base year 2008)	−12.4%	−20%	
Final energy consumption in transport (base year 2005)	1.7%	−10%	−40%

Recent progress toward improved efficiency has been steady aside from the financial crisis of 2007–2008. Some however believe energy efficiency is still under-recognised in terms of its contribution to Germany's energy transformation (or *Energiewende*).

Efforts to reduce final energy consumption in transport sector have not been successful, with a growth of 1.7% between 2005–2014. This growth is due to both road passenger and road freight transport. Both sectors increased their overall distance travelled to record the highest figures ever for Germany. Rebound effects played a significant role, both between improved vehicle efficiency and the distance travelled, and between improved vehicle efficiency and an increase in vehicle weights and engine power.

On 3 December 2014, the German federal government released its National Action Plan on Energy Efficiency (NAPE). The areas covered are the energy efficiency of buildings, energy conservation for companies, consumer energy efficiency, and transport energy efficiency. The policy contains both immediate and forward-looking measures. The central short-term measures of NAPE include the introduction of competitive tendering for energy efficiency, the raising of funding for building renovation, the introduction of tax incentives for efficiency measures in the building sector, and the setting up energy efficiency networks together with business and industry. German industry is expected to make a sizeable contribution.

On 12 August 2016, the German government released a green paper on energy efficiency for public consultation (in German). It outlines the potential challenges and actions needed to reduce energy consumption in Germany over the coming decades. At the document's launch, economics and energy minister Sigmar Gabriel said "we do not need to produce, store, transmit and pay for the energy that we save". The green paper prioritizes the efficient use of energy as the "first" response and also outlines opportunities for sector coupling, including using renewable power for heating and transport. Other proposals include a flexible energy tax which rises as petrol prices fall, thereby incentivizing fuel conservation despite low oil prices.

Poland

In May 2016 Poland adopted a new Act on Energy Efficiency, to enter into force on 1 October 2016.

United States

A 2011 Energy Modeling Forum study covering the United States examines how energy efficiency opportunities will shape future fuel and electricity demand over the next several decades. The US economy is already set to lower its energy and carbon intensity, but explicit policies will be necessary to meet climate goals. These policies include: a carbon tax, mandated standards for more efficient appliances, buildings and vehicles, and subsidies or reductions in the upfront costs of new more energy efficient equipment.

Industry

Industries use a large amount of energy to power a diverse range of manufacturing and resource extraction processes. Many industrial processes require large amounts of heat and mechanical power, most of which is delivered as natural gas, petroleum fuels and as electricity. In addition some industries generate fuel from waste products that can be used to provide additional energy.

Because industrial processes are so diverse it is impossible to describe the multitude of possible opportunities for energy efficiency in industry. Many depend on the specific technologies and processes in use at each industrial facility. There are, however, a number of processes and energy services that are widely used in many industries.

Various industries generate steam and electricity for subsequent use within their facilities. When electricity is generated, the heat that is produced as a by-product can be captured and used for process steam, heating or other industrial purposes. Conventional electricity generation is about 30% efficient, whereas combined heat and power (also called co-generation) converts up to 90 percent of the fuel into usable energy.

Advanced boilers and furnaces can operate at higher temperatures while burning less fuel. These technologies are more efficient and produce fewer pollutants.

Over 45 percent of the fuel used by US manufacturers is burnt to make steam. The typical industrial facility can reduce this energy usage 20 percent (according to the US Department of Energy) by insulating steam and condensate return lines, stopping steam leakage, and maintaining steam traps.

Electric motors usually run at a constant speed, but a variable speed drive allows the motor's energy output to match the required load. This achieves energy savings ranging from 3 to 60 percent, depending on how the motor is used. Motor coils made of superconducting materials can also reduce energy losses. Motors may also benefit from voltage optimisation.

Industry uses a large number of pumps and compressors of all shapes and sizes and in a wide variety of applications. The efficiency of pumps and compressors depends on many factors but often improvements can be made by implementing better process control and better maintenance practices. Compressors are commonly used to provide compressed air which is used for sand blasting, painting, and other power tools. According to the US Department of Energy, optimizing compressed air systems by installing variable speed drives, along with preventive maintenance to detect and fix air leaks, can improve energy efficiency 20 to 50 percent.

Transportation

Automobiles

Toyota Prius used by NYPD Traffic Enforcement

The estimated energy efficiency for an automobile is 280 Passenger-Mile/10^6 Btu. There are several ways to enhance a vehicle's energy efficiency. Using improved aerodynamics to minimize drag can increase vehicle fuel efficiency. Reducing vehicle weight can also improve fuel economy, which is why composite materials are widely used in car bodies.

More advanced tires, with decreased tire to road friction and rolling resistance, can save gasoline. Fuel economy can be improved by up to 3.3% by keeping tires inflated to the correct pressure. Replacing a clogged air filter can improve a cars fuel consumption by as much as 10 percent on older vehicles. On newer vehicles (1980s and up) with fuel-injected, computer-controlled engines, a clogged air filter has no effect on mpg but replacing it may improve acceleration by 6-11 percent.

Turbochargers can increase fuel efficiency by allowing a smaller displacement engine. The 'Engine of the year 2011' is a Fiat 500 engine equipped with an MHI turbocharger. "Compared with a 1.2-liter 8v engine, the new 85 HP turbo has 23% more power and a 30% better performance index. The performance of the two-cylinder is not only equivalent to a 1.4-liter 16v engine, but fuel consumption is 30% lower."

Energy-efficient vehicles may reach twice the fuel efficiency of the average automobile. Cutting-edge designs, such as the diesel Mercedes-Benz Bionic concept vehicle have achieved a fuel efficiency as high as 84 miles per US gallon (2.8 L/100 km; 101 mpg_{-imp}), four times the current conventional automotive average.

The mainstream trend in automotive efficiency is the rise of electric vehicles (all@electric or hybrid electric). Hybrids, like the Toyota Prius, use regenerative braking to recapture energy that would dissipate in normal cars; the effect is especially pronounced in city driving. Plug-in hybrids also have increased battery capacity, which makes it possible to drive for limited distances without burning any gasoline; in this case, energy efficiency is dictated by whatever process (such as coal-burning, hydroelectric, or renewable source) created the power. Plug-ins can typically drive for around 40 miles (64 km) purely on electricity without recharging; if the battery runs low, a gas engine kicks in allowing for extended range. Finally, all-electric cars are also growing in popularity; the Tesla Model S sedan is the only high-performance all-electric car currently on the market.

Street Lighting

Cities around the globe light up millions of streets with 300 million lights. Some cities are seeking to reduce street light power consumption by dimming lights during off-peak hours or switching to LED lamps. It is not clear whether the high luminous efficiency of LEDs will lead to real reductions in energy, as cities may end up installing extra lamps or lighting areas more brightly than in the past.

Aircraft

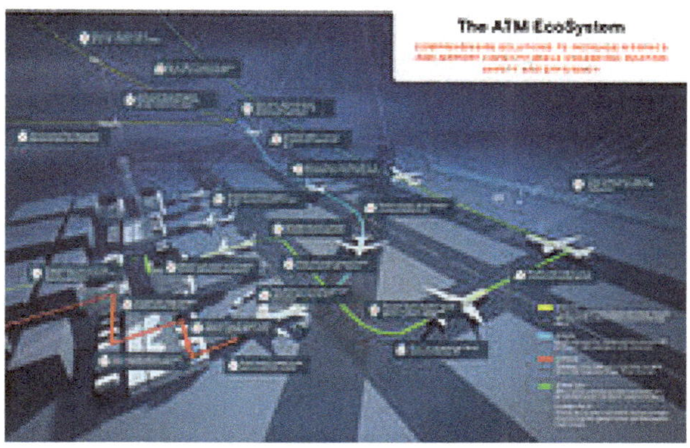

Air Traffic Management

There are several ways to reduce energy usage in air transportation, from modifications to the planes themselves, to how air traffic is managed. As in cars, turbochargers are an effective way to reduce energy consumption; however, instead of allowing for the use of a smaller-displacement engine, turbochargers in jet turbines operate by compressing the thinner air at higher altitudes. This allows the engine to operate as if it were at sea-level pressures while taking advantage of the reduced drag on the aircraft at higher altitudes.

Air traffic management systems are another way to increase the efficiency of not just the aircraft but the airline industry as a whole. New technology allows for superior automation of takeoff, landing, and collision avoidance, as well as within airports, from simple things like HVAC and lighting to more complex tasks such as security and scanning.

Alternative Fuels

Alternative fuels, known as non-conventional or advanced fuels, are any materials or substances that can be used as fuels, other than conventional fuels. Some well known alternative fuels include biodiesel, bioalcohol (methanol, ethanol, butanol), chemically stored electricity (batteries and fuel cells), hydrogen, non-fossil methane, non-fossil natural gas, vegetable oil, and other biomass sources.

Typical Brazilian filling station with four alternative fuels for sale: biodiesel (B3), gasohol (E25), neat ethanol (E100), and compressed natural gas (CNG). Piracicaba, Brazil.

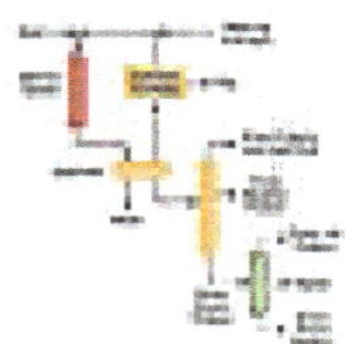

Green diesel production

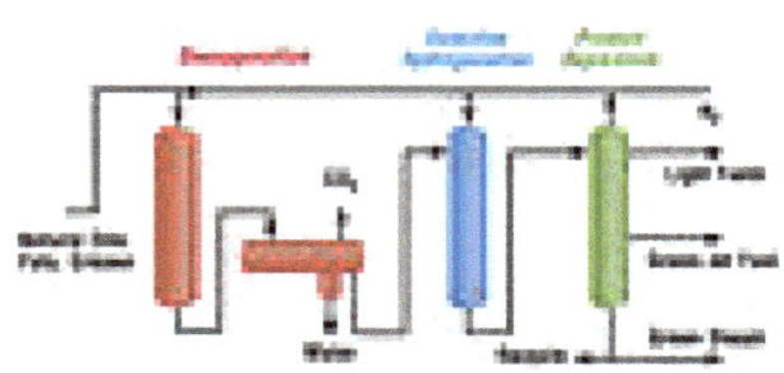

Green jet fuel production

Energy Conservation

Energy conservation is broader than energy efficiency in including active efforts to decrease energy consumption, for example through behaviour change, in addition to using energy more efficiently. Examples of conservation without efficiency improvements are heating a room less in winter, using the car less, air-drying your clothes instead of using the dryer, or enabling energy saving modes on a computer. As with other definitions, the boundary between efficient energy use and energy conservation can be fuzzy, but both are important in environmental and economic terms. This is especially the case when actions are directed at the saving of fossil fuels. Energy conservation is a challenge requiring policy programmes, technological development and behavior change to go hand in hand. Many energy intermediary organisations, for example governmental or non-governmental organisations on local, regional, or national level, are working on often publicly funded programmes or projects to meet this challenge. Psychologists have also engaged with the issue of energy conservation and have provided guidelines for realizing behavior change to reduce energy consumption while taking technological and policy considerations into account.

The National Renewable Energy Laboratory maintains a comprehensive list of apps useful for energy efficiency.

Commercial property managers that plan and manage energy efficiency projects generally use a software platform to perform energy audits and to collaborate with contractors to understand their full range of options. The Department of Energy (DOE) Software Directory describes EnergyActio software, a cloud based platform designed for this purpose.

Sustainable Energy

Energy efficiency and renewable energy are said to be the "twin pillars" of a sustainable energy policy. Both strategies must be developed concurrently in order to stabilize and reduce carbon dioxide emissions. Efficient energy use is essential to slowing the energy demand growth so that rising clean energy supplies can make deep cuts in fossil fuel use. If energy use grows too rapidly, renewable energy development will chase a receding target. Likewise, unless clean energy supplies come online rapidly, slowing demand growth will only begin to reduce total carbon emissions; a reduction in the carbon content of energy sources is also needed. A sustainable energy economy thus requires major commitments to both efficiency and renewables.

Rebound Effect

If the demand for energy services remains constant, improving energy efficiency will reduce energy consumption and carbon emissions. However, many efficiency improvements do not reduce energy consumption by the amount predicted by simple engineering models. This is because they make energy services cheaper, and so consumption of those services increases. For example, since fuel efficient vehicles make travel cheaper, consumers may choose to drive farther, thereby offsetting some of the potential energy savings. Similarly, an extensive historical analysis of technological efficiency improvements has conclusively shown that energy efficiency improvements were almost always outpaced by economic growth, resulting in a net increase in resource use and associated pollution. These are examples of the direct rebound effect.

Estimates of the size of the rebound effect range from roughly 5% to 40%. The rebound effect is likely to be less than 30% at the household level and may be closer to 10% for transport. A rebound effect of 30% implies that improvements in energy efficiency should achieve 70% of the reduction in energy consumption projected using engineering models. The rebound effect may be particularly large for lighting, because in contrast to tasks like transport there is effectively no upper limit on how much light could be considered useful. In fact, it appears that lighting has accounted for about 0.7% of GDP across many societies and hundreds of years, implying a rebound effect of 100%.

Energy Efficiency in Transportation

Velomobile, the most efficient means of passenger transportation

Energy efficiency in transportation is, strictly speaking, useful output (kilometers, miles, etc.) divided by total energy input (kilowatt-hour, amount of gasoline, etc.). This is more precisely known as fuel economy (symbol FE), and may be measured in units of km/L or mi/gal (MPG).

Energy efficiency in transportation is often (and confusingly) described in terms of fuel consumption (symbol FC). Fuel consumption is the reciprocal of fuel economy and is occasionally known as energy intensity; it equals fuel (energy) divided by distance, and is often measured in units of liters per 100 kilometers (L/100 km), joules per kilometer (J/km), gallon-equivalent per 100 miles (gal_e/100 mi), or kilowatt-hours per distance (kW-hr/km or kW-hr/mi).

The energy efficiency in transportation varies by mode of transportation (listed below). Different types of transportation range from some hundred kilojoules per kilometre (kJ/km) for a bicycle to tens of megajoules per kilometer (MJ/km) for a helicopter.

Units of Measurement

Both fuel economy and fuel consumption are often given on a per-vehicle basis. More refined units of measurement exist, such as fuel consumption per passenger or fuel consumption per mass of cargo.

Energy efficiency is expressed in terms of fuel economy:

- distance per vehicle per unit fuel volume; e.g., km/L or miles per gallon (US or imperial).

- distance per vehicle per unit fuel mass; e.g., km/kg.
- distance per vehicle per unit energy; e.g., miles per gallon equivalent (mpg-e).

Energy intensity (reciprocal efficiency) is expressed terms of fuel consumption:

- volume of fuel (or total energy) consumed per unit distance per vehicle; e.g. L/100 km or MJ/100 km.
- volume of fuel (or total energy) consumed per unit distance per passenger; e.g., L/(100 passenger·km).
- volume of fuel (or total energy) consumed per unit distance per unit mass of cargo transported; e.g., L/100 kg·km or MJ/t·km.
- electricity used per vehicle per unit distance; e.g., kW·h/100 km.

Transportation Types

For freight transport, rail and ship transport are generally much more efficient than trucking, and air freight is much less efficient.

Walking

Nordic walkers

A 68 kg (150 lb) person walking at 4 km/h (2.5 mph) requires approximately 210 kilocalories (880 kJ) of food energy per hour, which is equivalent to 4.55 km/MJ. 1 US gal (3.8 L) of gasoline contains about 114,000 British thermal units (120 MJ) of energy, so this is approximately equivalent to 360 miles per US gallon (0.65 L/100 km).

Velomobile

Velomobiles seem to have the highest energy efficiency in personal transportation. At a speed of 50 km/h (31 mph) the WAW manufacturer claims they need only 0.5 kW·h of food energy per 100 km to transport the passenger, which is around $\frac{1}{5}$ (20%) of a normal bicycle, and $1/_{50}$ (2%) of an average fossil fuel or electric car. This corresponds to 4700 miles per US gallon (2000 km/L, 0.05 L/100 km). Other sources give a figure of $1/_{3.4}$ (29.5%) of the energy use of a normal bicycle.

Bicycling

A Chinese Flying Pigeon bicycle

A relatively light and slow vehicle with low-friction tires and an efficient chain-driven drivetrain, the bicycle is one of the most energy-efficient forms of transport. Compared with walking, a 64 kg (140 lb) cyclist riding at 16 km/h (10 mph) requires about half the food energy per unit distance: 43 kcal/mi, 27 kcal/km or 3.1 kW·h (11 MJ) per 100 km. This converts to about 732 mpg_{-US} (0.321 L/100 km; 879 mpg_{-imp}). This figure depends on the speed and mass of the rider: greater speeds give higher air drag and heavier riders consume more energy per unit distance.

A motorized bicycle allows human power and the assistance of a 49 cm^3 (3.0 cu in) engine, giving a range of 160 to 200 mpg_{-US} (1.5–1.2 L/100 km; 190–240 mpg_{-imp}). Electric pedal-assisted bikes run on as little as 1.0 kW·h (3.6 MJ) per 100 km, while maintaining speeds in excess of 30 km/h (19 mph). These best-case figures rely on a human doing 70% of the work, with around 3.6 MJ (1.0 kW·h) per 100 km coming from the motor.

Human Power

To be thorough, a comparison must also consider the energy costs of producing, transporting and packaging of fuel (food or fossil fuel), the energy incurred in disposing of exhaust waste, and the energy costs of manufacturing the vehicle. This last can be significant given that walking requires little or no special equipment, while automobiles, for example, require a great deal of energy to produce and have relatively short lifespans. In addition, any comparison of electric vehicles and liquid-fuelled vehicles must include the fuel consumed in the power station to generate the electricity. In the UK for instance the efficiency of the electricity generation and distribution system is around 0.40.

Automobiles

Bugatti Veyron

Automobile fuel efficiency is most commonly expressed in terms of the volume of fuel consumed per one hundred kilometres (L/100 km), but in some countries (including the USA, UK and India) it is more commonly expressed in terms of the distance per volume fuel consumed (km/L or miles per US or imperial gallon). This is complicated by the different energy content of fuels such as petrol and diesel. The Oak Ridge National Laboratory (ORNL) states that the energy content of unleaded gasoline is 115,000 British thermal unit (BTU) per US gallon (32 MJ/L) compared to 130,500 BTU per US gallon (36.4 MJ/L) for diesel.

A second important consideration is the energy costs of producing energy. Bio-fuels, electricity and hydrogen, for instance, have significant energy inputs in their production. Because of this, the 50–70% efficiency of hydrogen production has to be combined with the vehicle efficiency to yield net efficiency.

A third consideration to take into account is the occupancy rate of the vehicle. As the number of passengers per vehicle increases the consumption per unit distance per vehicle increases. However this increase is slight compared to the reduction in consumption per unit distance per passenger. We can compare, for instance, the estimated average occupancy rate of about 1.3 passengers per car in the San Francisco Bay Area to the 2006 UK estimated average of 1.58.

Finally, vehicle energy efficiency calculations would be misleading without factoring the energy cost of producing the vehicle itself. This initial energy cost can of course be depreciated over the life of the vehicle to calculate an average energy efficiency over its effective life span. In other words, vehicles that take a lot of energy to produce and are used for relatively short periods will require a great deal more energy over their effective lifespan than those that do not, and are therefore much less energy efficient than they may otherwise seem. Compare, for example, walking, which requires no special equipment at all, and an automobile, produced in and shipped from another country, and made from parts manufactured around the world from raw materials and minerals mined and processed elsewhere again, and used for a limited number of years.

Driving practices and vehicles can be modified to improve their energy efficiency by about 15%.

Example Consumption Figures

Two American solar cars in Canada

- Solar cars use no fuel, charging the batteries entirely from built-in solar panels, and typically use less than 3 kW·h per 100 miles (67 kJ/km or 1.86 kW·h/100 km).
- The four passenger GEM NER uses 169 Wh/mi (203 mpg-e; 10.5 kW·h/100 km), which equates to 2.6 kW·h/100 km per person when fully occupied, albeit at only 24 mph (39 km/h).
- The General Motors EV1 was rated in a test with a charging efficiency of 373 Wh-AC/mile or 23 kWh/100 km approximately equivalent to 2.6 L/100 km (110 mpg_{-imp}; 90 mpg_{-US}) for petroleum-fuelled vehicles.
- Chevrolet Volt in full electric mode uses 36 kilowatt-hours per 100 miles (810 kJ/km; 96 mpg-e), meaning it may be more energy-efficient than walking for 4 or more passengers.
- The Daihatsu Charade 993cc turbo diesel (1987–1993) won the most fuel efficient vehicle award for going round the United Kingdom consuming an average of 2.82 L/100 km (100 mpg_{-imp}). It was surpassed only recently by the VW Lupo 3 L which consumes about 2.77 L/100 km (102 mpg_{-imp}). Both cars are rare to find on the popular market. The Daihatsu had major problems with rust and structural safety which contributes to its rarity and the quite short production run.
- The Volkswagen Polo 1.4 TDI Bluemotion and the SEAT Ibiza 1.4 TDI Ecomotion, both rated at 3.8 L/100 km (74 mpg_{-imp}; 62 mpg_{-US}) (combined) were the most fuel efficient cars on sale in the UK as of 22 March 2008.
- Honda Insight – achieves 48 mpg_{-US} (4.9 L/100 km; 58 mpg_{-imp}) under real-world conditions.

- Honda Civic Hybrid- regularly averages around 45 mpg_{-US} (5.2 L/100 km; 54 mpg_{-imp}).
- Toyota Prius – According to the US EPA's revised estimates, the combined fuel consumption for the 2008 Prius is 46 mpg_{-US} (5.1 L/100 km; 55 mpg_{-imp}), making it the most fuel efficient United States car of 2008. In the UK, the official fuel consumption figure (combined) for the Prius is 4.3 L/100 km (66 mpg_{-imp}; 55 mpg_{-US}).
- 2012 Cadillac CTS-V Wagon 6.2 L Supercharged, 14 mpg_{-US} (17 L/100 km; 17 mpg_{-imp}).
- 2012 Bugatti Veyron, 10 mpg_{-US} (24 L/100 km; 12 mpg_{-imp}).

Aircraft

A principal determinant of energy consumption in aircraft is drag, which must be opposed by thrust for the aircraft to progress.

- Drag is proportional to the lift required for flight, which is equal to the weight of the aircraft. As induced drag increases with weight, mass reduction, with improvements in engine efficiency and reductions in aerodynamic drag, has been a principal source of efficiency gains in aircraft, with a rule-of-thumb being that a 1% weight reduction corresponds to around a 0.75% reduction in fuel consumption.
- Flight altitude affects engine efficiency. Jet-engine efficiency increases at altitude up to the tropopause, the temperature minimum of the atmosphere; at lower temperatures, the Carnot efficiency is higher. Jet engine efficiency is also increased at high speeds, but above about Mach 0.85 the airframe aerodynamic losses increase faster.
- Compressibility effects: beginning at transonic speeds of around Mach 0.85, shockwaves form increasing drag.
- For supersonic flight, it is difficult to achieve a lift to drag ratio greater than 5, and fuel consumption is increased in proportion.

Concorde fuel efficiency comparison (assuming jets are filled to capacity)		
Aircraft	**Concorde**	**Boeing 747-400**
Passenger-miles/imperial gallon	17	109
Passenger-miles/US gallon	14	91
Litres/100 passenger-km	16.6	3.1

Passenger airplanes averaged 4.8 l/100 km per passenger (1.4 MJ/passenger-km) (49 passenger-miles per gallon) in 1998. Note that on average 20% of seats are left unoccupied. Jet aircraft efficiencies are improving: Between 1960 and 2000 there was a 55% overall fuel efficiency gain (if one were to exclude the inefficient and limited fleet of the DH Comet 4 and to consider the Boeing 707 as the base case). Most of the improvements in efficiency were gained in the first decade when jet craft first came into widespread commercial use. Compared to advanced piston engine airliners of the 1950s, current jet airliners are only marginally more efficient per passenger-mile. Between 1971 and 1998 the fleet-average annual improvement per available seat-kilometre was estimated at 2.4%. Concorde the supersonic transport managed about 17 passenger-miles to the Imperial gallon; similar to a business jet, but much worse than a subsonic turbofan aircraft. Airbus puts the fuel rate consumption of their A380 at less than 3 l/100 km per passenger (78 passenger-miles per US gallon).

Air France Airbus A380-800

The mass of an aircraft can be reduced by using light-weight materials such as titanium, carbon fiber and other composite plastics. Expensive materials may be used, if the reduction of mass justifies the price of materials through improved fuel efficiency. The improvements achieved in fuel efficiency by mass reduction, reduces the amount of fuel that needs to be carried. This further reduces the mass of the aircraft and therefore enables further gains in fuel efficiency. For example, the Airbus A380 design includes multiple light-weight materials.

Airbus has showcased wingtip devices (sharklets or winglets) that can achieve 3.5 percent reduction in fuel consumption. There are wingtip devices on the Airbus A380. Further developed Minix winglets have been said to offer 6 percent reduction in fuel consumption. Winglets at the tip of an aircraft wing smooth out the wing-tip vortex (reducing the aircraft's wing drag) and can be retrofitted to any airplane.

NASA and Boeing are conducting tests on a 500 lb (230 kg) "blended wing" aircraft. This design allows for greater fuel efficiency since the whole craft produces lift, not just the wings. The blended wing body (BWB) concept offers advantages in structural, aerodynamic and operating efficiencies over today's more conventional fuselage-and-wing designs. These features translate into greater range, fuel economy, reliability and life cycle savings, as well as lower manufacturing costs. NASA has created a cruise efficient STOL (CESTOL) concept.

Fraunhofer Institute for Manufacturing Engineering and Applied Materials Research (IFAM) have researched a shark skin imitating paint that would reduce drag through a riblet effect. Aircraft are a major potential application for new technologies such as aluminium metal foam and nanotechnology such as the shark skin imitating paint.

Propfan propulsors are a more fuel efficient technology than jets or turboprops, but turboprops have an optimum speed below about 450 mph (700 km/h). This speed is less than used with jets by major airlines today. However, the decrease in speed reduces drag. With the current high price for jet fuel and the emphasis on engine/airframe efficiency to reduce emissions, there is renewed interest in the propfan concept for jetliners that might come into service beyond the Boeing 787 and Airbus A350XWB. For instance, Airbus has patented aircraft designs with twin rear-mounted counter-rotating propfans. NASA has conducted an Advanced Turboprop Project (ATP), where they researched a variable pitch propfan that produced less noise and achieved high speeds.

Related to fuel efficiency is the impact of aviation emissions on climate.

Small Aircraft

Dyn'Aéro MCR4S

- Motor-gliders made of composite materials can reach an extremely low fuel consumption for cross-country flights, using thermal air currents and winds.
- At 160 km/h, a diesel powered two-seater Dieselis burns 6 liters of fuel per hour, 1.9 liters per 100 passenger km.
- at 220 km/h, a four-seater 100 hp MCR-4S burns 20 liters of gas per hour, 2.2 liters per 100 passenger km.
- Under continuous motorised flight at 225 km/h, a Pipistrel Sinus burns 11 liters of fuel per flight hour. Carrying 2 people aboard, it operates at 2.4 liters per 100 passenger km.
- Ultralight aircraft Tecnam P92 Echo Classic at cruise speed of 185 km/h burns 17 liters of fuel per flight hour, 4.6 liters per 100 passenger km (2 people). Other

modern ultralight aircraft have increased efficiency; Tecnam P2002 Sierra RG at cruise speed of 237 km/h burns 17 liters of fuel per flight hour, 3.6 liters per 100 passenger km (2 people).

- Two-seater and four-seater flying at 250 km/h with old generation engines can burn 25 to 40 liters per flight hour, 3 to 5 liters per 100 passenger km.
- The Sikorsky S-76C++ twin turbine helicopter gets about 1.65 mpg_{-US} (143 L/100 km; 1.98 mpg_{-imp}) at 140 knots (260 km/h; 160 mph) and carries 12 for about 19.8 passenger-miles per gallon (11.9 L per 100 passenger km).

Ships

Queen Elizabeth

RMS *Queen Elizabeth 2*

Cunard states that RMS *Queen Elizabeth 2* travels 49.5 feet per imperial gallon of diesel oil (3.32 m/l or 41.2 ft/US gal), and that it has a passenger capacity of 1777. Thus carrying 1777 passengers we can calculate an efficiency of 16.7 passenger miles per imperial gallon (16.9 l/100 p·km or 13.9 p·mpg_{-US}).

Cruise Ships

MS *Oasis of the Seas* has a capacity of 6,296 passengers and a fuel efficiency of 14.4 passenger miles per US gallon. Voyager-class cruise ships have a capacity of 3,114 passengers and a fuel efficiency of 12.8 passenger miles per US gallon.

Emma Maersk

Emma Maersk uses a Wärtsilä-Sulzer RTA96-C, which consumes 163 g/kW·h and 13,000 kg/h. If it carries 13,000 containers then 1 kg fuel transports one container for one hour over a distance of 45 km. The ship takes 18 days from Tanjung (Singapore) to Rotterdam (Netherlands), 11 from Tanjung to Suez, and 7 from Suez to Rotterdam, which is roughly 430 hours, and has 80 MW, +30 MW. 18 days at a mean speed of 25 knots (46 km/h) gives a total distance of 10,800 nautical miles (20,000 km).

Sailboats

A sailboat, much like a solar car, can locomote without consuming any fuel. A sail boat such as a Dinghy using just wind power requires no input energy in terms of fuel. However some manual energy is required by the crew to steer the boat and adjust the sails using ropes. In addition energy will be needed for demands other than propulsion, such as cooking, heating or lighting. The fuel efficiency of a single-occupancy boat is highly dependent on the size of its engine, the speed at which it travels, and its displacement. Due to the high viscosity of water, with a single passenger, the equivalent energy efficiency will be lower than in a car, train, or plane.

Trains

Trains can be an efficient means of transport for freight and passengers. Efficiency varies significantly with passenger loads, and losses incurred in electricity generation and supply (for electrified systems), and, importantly, end-to-end delivery, where stations are not the originating final destinations of a journey.

Actual consumption depends on gradients, maximum speeds, loading and stopping patterns. Data produced for the European MEET project (Methodologies for Estimating Air Pollutant Emissions) illustrate the different consumption patterns over several track sections. The results show the consumption for a German ICE high-speed train varied from around 19 to 33 kW·h/km (68–119 MJ/km; 31–53 kW·h/mi). The data also reflects the weight of the train per passenger. For example, TGV double-deck Duplex trains use lightweight materials, which keep axle loads down and reduce damage to track and also save energy.

Freight

Energy consumption estimates for rail freight vary widely, and many are provided by interested parties. Some are tabulated below.

Country	Year	Fuel economy (weight of goods)	Energy Intensity
USA	2007	185.363 km/L (1 short ton)	*energy/mass-distance*
UK	—	87 t·km/L	0.41 MJ/t·km (LHV)

Passenger

Country	Year	Train efficiency	Per passenger-km (kJ)
Japan	2004	17.9 MJ (5.0 kWh)/car-km	350 kJ/passenger-km
EC	1997	18 kW·h/km (65 MJ/km)	

USA		1.125 mpg_{-US} (209.1 L/100 km; 1.351 mpg_{-imp})	468 passenger-miles/US gallon (0.503 L/100 passenger-km)
Switzerland	2011	2300 GWhr/yr	470 kJ/passenger-km
Basel, Switzer-land		1.53 kWh/vehicle-km (5.51 MJ/vehicle-km)	85 kJ/passenger-km (150 kJ/passenger-km at 80% average load)
USA	2009		2,435 BTU/mi (1.60 MJ/km)
Portugal	2011	8.5 kW·h/km (31 MJ/km; 13.7 kW·h/mi)	77 kJ/passenger-km

Considering only the energy spent to move the train, and taking as example the urban area of Lisbon, train seems to be on average 20 times more efficient than automobile for transportation of passengers, if we consider energy spent per passenger-km. Considering an automobile which has a consumptions of around 6 l/100 km (47 mpg_{-imp}; 39 mpg_{-US}) of gasoline, the fact the on average cars in Europe have an occupation ratio of around 1.2 passengers per automobile and that one litre of gasoline amounts for about 8826 Wh, one gets on average 441 Wh (1,590 kJ) per passenger-km. On the other hand, a modern urban train with an average occupation of 20% of total capacity, which has a consumption of about 8.5 kW·h/km (31 MJ/km; 13.7 kW·h/mi), one gets 21.5 Wh per passenger-km, 20 times less than the automobile.

Braking Losses

Stopping is a considerable source of inefficiency. Modern electric trains like the *Shinkansen* (the *Bullet Train*) use regenerative braking to return current into the catenary while they brake. A Siemens study indicated that regenerative braking might recover 41.6% of the total energy consumed. The Passenger Rail (Urban and Intercity) and Scheduled Intercity and All Charter Bus Industries Technological and Operational Improvements – FINAL REPORT states that "Commuter operations can dissipate more than half of their total traction energy in braking for stops." and that "We estimate head-end power to be 35 percent (but it could possibly be as high as 45 percent) of total energy consumed by commuter railways." Having to accelerate and decelerate a heavy train load of people at every stop is inefficient despite regenerative braking which can recover typically around 20% of the energy wasted in braking. Weight is a determinant of braking losses.

Other References

AEA study of road and rail for the United Kingdom Department for Transport: Final report

Buses

The Bus Rapid Transit of Metz uses a diesel-electric hybrid driving system, developed by Belgian Van Hool manufacturer.

- In July 2005, the average occupancy for buses in the UK was stated to be 9.
- The fleet of 244 40-foot (12 m) 1982 New Flyer trolley buses in local service with BC Transit in Vancouver, Canada, in 1994/95 used 35,454,170 kWh for 12,966,285 vehicle km, or 9.84 MJ/vehicle km. Exact ridership on trolleybuses is not known, but with all 34 seats filled this equates to 0.32 MJ/passenger km. It is quite common to see people standing on Vancouver trolleybuses. This is a service with many stops per kilometre; part of the reason for the efficiency is the use of regenerative braking.
- A commuter service in Santa Barbara, California, USA, found average diesel bus efficiency of 6.0 mpg_{-US} (39 L/100 km; 7.2 mpg_{-imp}) (using MCI 102DL3 buses). With all 55 seats filled this equates to 330 passenger mpg; with 70% filled, 231 passenger mpg.
- In 2011 the fleet of 752 buses in the city of Lisbon had an average speed of 14.4 km/h and an average occupancy of 20.1 passengers per vehicle.

Other

- NASA's Crawler-Transporter is used to move the Shuttle from storage to the launch pad. It uses diesel and has one of the highest fuel consumption rates on record, 150 US gallons per mile (350 l/km; 120 imp gal/mi).

International Transport Comparisons

UK Public transport

Rail and bus are generally required to serve 'off peak' and rural services, which by their nature have lower loads than city bus routes and inter city train lines. Moreover, due to their 'walk on' ticketing it is much harder to match daily demand and passenger numbers. As a consequence, the overall load factor on UK railways is 35% or 90 people per train:

Conversely, Air services work on point-to-point networks between large population centres and are 'pre-book' in nature. Using yield management overall loads can be raised to around 70–90%. However, recently intercity train operators have been using similar techniques, with loads reaching typically 71% overall for TGV services in France and a similar figure for the UK's Virgin Trains services.

For emissions, the electricity generating source needs to be taken into account.

US Passenger Transportation

The US Transportation Energy Data Book states the following figures for passenger transportation in 2009:

Transport mode	Average passengers per vehicle	BTU per passenger-mile	MJ per passenger-kilometre
Rail (intercity Amtrak)	20.9	2,435	1.596
Motorcycles	1.16	2,460	1.61
Rail (transit light & heavy)	24.5	2,516	1.649
Rail (commuter)	32.7	2,812	1.843
Air	99.3	2,826	1.853
Cars	1.55	3,538	2.319
Personal trucks	1.84	3,663	2.401
Buses (transit)	9.2	4,242	2.781
Taxi	1.55	15,645	10.257

US Freight transportation

The US Transportation Energy book states the following figures for freight transportation in 2010:

Transportation mode	Fuel consumption	
	BTU per short ton-mile	kJ per tonne-kilometre
Domestic waterborne	217	160
Class 1 railroads	289	209
Heavy trucks	3,357	2,426
Air freight (approx.)	9,600	6,900

From 1960 to 2010 the efficiency of air freight has increased 75%, mostly due to more efficient jet engines.

1 gal_{-US} (3.785 l, 0.833 gal_{-imp}) of fuel can move a ton of cargo 857 km or 462 nmi by barge, or 337 km (209 mi) by rail, or 98 km (61 mi) by truck.

Compare:

- Space Shuttle used to transport freight to the other side of the Earth: 40 megajoules per tonne-kilometre.
- Net energy for lifting: 10 megajoules per tonne-kilometre.

Canadian Transportation

Natural Resources Canada's Office of Energy Efficiency publishes annual statistics regarding the efficiency of the entire Canadian fleet. For researchers, these fuel consumption estimates are more realistic than the fuel consumption ratings of new vehicles, as they represent the real world driving conditions, including extreme weather and traffic. The annual report is called Energy Efficiency Trends Analysis. There are dozens of tables illustrating trends in energy consumption expressed in energy per passenger km (passengers) or energy per tonne km (freight).

French Environmental Calculator

The environmental calculator of the French environment and energy agency (ADEME) published in 2007 using data from 2005 enables one to compare the different means of transportation as regards the CO_2 emissions (in terms of carbon dioxide equivalent) as well as the consumption of primary energy. In the case of an electric vehicle, the ADEME makes the assumption that 2.58 toe as primary energy are necessary for producing one toe of electricity as end energy in France.

This computer tool devised by the ADEME shows the importance of public transportation from an environmental point of view. It highlights the primary energy consumption as well as the CO_2 emissions due to transportation. However, nothing is said about the production of radioactive waste. Moreover, intermodal passenger transport is probably a key to sustainable transport, by allowing people to use less polluting means of transportation.

Energy Efficiency Implementation

Energy efficiency implementation is an industry comprising firms which retrofit or replace inefficient equipment with more efficient parts or equipment, with the goal of reducing energy consumption. Energy efficiency implementation firms have the resources, both physical and intellectual, that can provide advice and materials to entities/facilities, assisting in decreasing energy use. Energy efficiency implementation is performed in two distinct ways: by retrofitting existing equipment or by replacing existing equipment completely. Retrofitting enhances existing equipment by making it expend less energy. The goal of this kind of implementation is to save kilowatts (kW) and kilowatt hours (kWh). The difference between these two measurements is that one is a power rating (kW) and the other is a measurement of energy actually consumed (kWh).

Utilities

Public Utility Commissions in many states mandate that their utilities design and implement energy efficiency programs. The funding for these can be reflected in their rates or are collected through a surcharge in monthly customer bills. Some utilities design their own programs and implement them using a rebate form or other application, and administer the programs using their own staff. Most major utilities hire implementation contractors who are responsible for the design and implementation, and some implement programs already designed and approved by their PUC. Some programs require a co-pay by the customer, some are installed at no-cost.

Utilities invest in energy efficiency for the following reasons:

1. Social interests: environmental issues and resource conservation
2. Economic interests: to prevent costly construction of power plants or the purchase of more energy

The end result is that utility companies have more energy to sell, or in other words, they are able to sell their excess capacity to more customers in the area without increasing their production capacity.

Effects

The results of energy efficiency implementation are all beneficial for the energy consumer. It reduces operational costs, reduces carbon footprint, and it can even improve quality of life. Energy efficiency implementation can also play a role in increased revenue when consumers choose a "greener" product over another that is not. Energy efficiency implementation can be extremely beneficial to large market segments like small businesses, schools, cement processing plants; basically any area that uses large amounts of energy. Small changes here add up to large savings.

Implementing energy efficiency measures in a home or business can also lead to behavioral changes. When an energy efficiency change has been made and the energy consumer sees the benefits, more changes are often wanted to further their savings. These small changes create awareness and can be as simple as turning of lights when a room is not in use, or as complex as adding window glazing or installing demand-control ventilation.

Role of Regulators

Energy sector regulators might have wide discretion in the implementation and/

or monitoring energy efficiency (EE) initiatives. The most likely roles involve giving technical advice to the agency developing EE initiatives, since changes in demand patterns will have implications for the operations and investment plans of utilities (and for costs, security of supply and quality of service) . Particularly when the EE outlays are by the utility, the energy sector regulator needs to monitor outcomes to ensure that the resources are being utilized in ways that are consistent with overarching public policies. Furthermore, interactions of utility initiatives with other EE policies need to be taken into account when evaluating whether the scale and scope of existing utility-based demand-side management programs. Utilities are in a position to analyze bills and conduct on-premises energy audits to identify areas of saving. Regulators could require utilities to undertake costly audit programs. A high tech approach to improving operations and the customer interface involves smart meters and information systems that enable the utility to track system performance in real time.

The costs of implementing such systems need to be balanced against the benefits, including the possibility that outlays on other projects might be more cost effective. Thus, the role of regulators primarily involves providing technical input into the development of EE policies initiated by other agencies or via legislated tax programs. In addition, the Regulator must determine (unless specified in law) which benefit-cost test is appropriate for evaluating utility-based EE programs. The regulatory tests include the participant cost test (will participants benefit over the measure's life?), the program administrator cost test (will utility bills increase?), the ratepayer impact measure (will utility prices increase?), the total resource cost test (will the total costs of energy decrease?) and the societal cost test (is the utility, state or nation better off, including environmental impacts?). The IEA Energy Efficiency Governance Handbook goes into much more detail on the importance of having a coherent system for developing, incentivizing, and evaluating energy efficiency programs.

Benefits

Operational Costs

By making appliances, lighting or the HVAC system more efficient, energy consumption levels drop, which leads to a lower utility bill for utility customers.

Carbon Footprint

By reducing energy consumption we are able to conserve natural resources and thus have a less dramatic impact on our environment.

Quality of Life

Working in natural or well-lit areas contributes to performance and productivity [citation here]. Comfortable temperature levels also factor into how well a person functions.

This fact can be applied to any operation that requires the use of lighting or an HVAC system.

Implementers

Energy efficiency projects can be implemented by commercial property managers or by energy service companies or contractors.

Commercial property managers that plan and manage energy efficiency projects generally use a software platform to perform energy audits and to collaborate with contractors to understand their full range of options.

Implementation companies normally focus on specific equipment or appliances that they specialize in retrofitting or replacing, or they provide this service by sector: residential, commercial or industrial. Energy efficiency implementation is a complex field, and in order to implement effectively, the implementer must be multi-faceted and have extensive experience in many areas of energy efficiency.

There are many systems, machines and methods that assist in creating energy savings including: gas, electricity, HVAC, lighting, daylighting, motion detection, insulated glazing, refrigerator strip curtains, revolving doors, anterooms, thermostat controls, demand-control ventilation and voltage optimisation

Energy Intensity

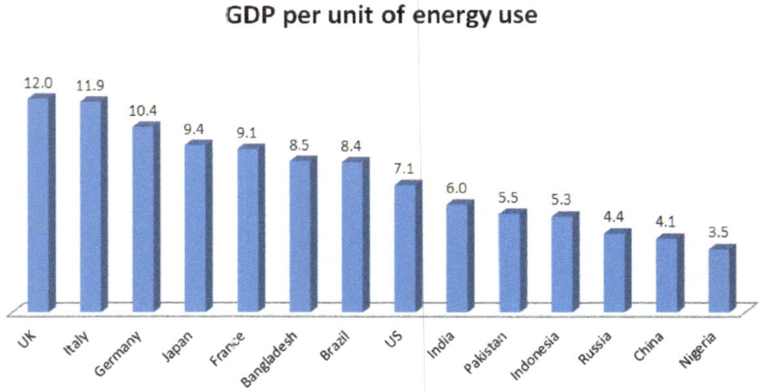

The World Bank : PPP $ per kg of oil equivalent (2011)

Energy intensity is a measure of the energy efficiency of a nation's economy. It is calculated as units of energy per unit of GDP.

- High energy intensities indicate a high price or cost of converting energy into GDP.

- Low energy intensity indicates a lower price or cost of converting energy into GDP.

High energy intensity means high industrial output as portion of GDP. Countries with low energy intensity signifies labor intensive economy,

> Energy Intensity as defined here is not to be confused with Energy Use Intensity (EUI), a measure of building energy use per unit area.

Overview

Many factors influence an economy's overall energy intensity. It may reflect requirements for general standards of living and weather conditions in an economy. It is not atypical for particularly cold or hot climates to require greater energy consumption in homes and workplaces for heating (furnaces, or electric heaters) or cooling (air conditioning, fans, refrigeration). A country with an advanced standard of living is more likely to have a wider prevalence of such consumer goods and thereby be impacted in its energy intensity than one with a lower standard of living.

Energy efficiency of appliances and buildings (through use of building materials and methods, such as insulation), fuel economy of vehicles, vehicular distances travelled (frequency of travel or larger geographical distances), better methods and patterns of transportation, capacities and utility of mass transit, energy rationing or conservation efforts, 'off-grid' energy sources, and stochastic economic shocks such as disruptions of energy due to natural disasters, wars, massive power outages, unexpected new sources, efficient uses of energy or energy subsidies may all impact overall energy intensity of a nation.

Thus, a nation that is highly economically productive, with mild and temperate weather, demographic patterns of work places close to home, and uses fuel efficient vehicles, supports carpools, mass transportation or walks or rides bicycles, will have a far lower energy intensity than a nation that is economically unproductive, with extreme weather conditions requiring heating or cooling, long commutes, and extensive use of generally poor fuel economy vehicles. Paradoxically, some activities that may seem to promote high energy intensities, such as long commutes, could in fact result in lower energy intensities by causing a disproportionate increase in GDP output.

Figures of energy consumption used in statistics are energy sources marketed through major energy industries. Therefore, some small scale but frequent consumption of energy source like firewood, charcoal peat, water wheel, wind mill are not in its count. In countries, which does not have such developed energy industries or people with highly self energy efficient life style, report smaller energy consumption figures.

Examples

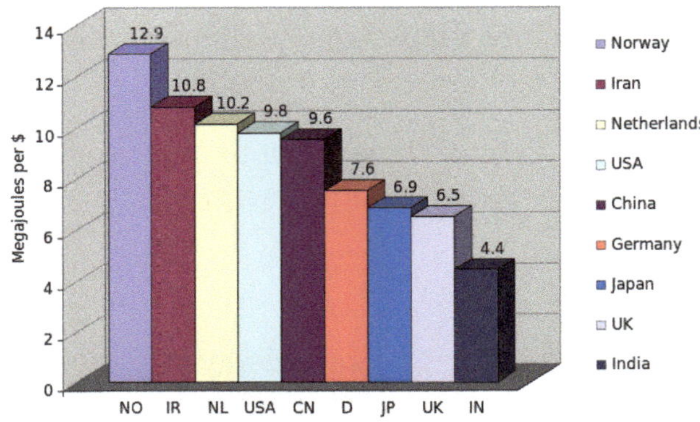

Energy Intensity of different economies The graph shows the amount of energy it takes to produce a US $ of GNP for selected countries. GNP is based on 2004 purchasing power parity and 2000 dollars adjusted for inflation.

- U.S. energy consumption from all sources in 2004 was estimated at 99.74 quad (105,210 PJ). Total GDP that year was $11.75 trillion, with per-capita GDP at $40,100. Using a population of 290,809,777, this would produce an Energy Intensity of 8,553 BTU (9,024 kJ) per dollar.

Various nations have significantly higher or lower energy intensities.

- Bangladesh, with a population of 144 million and a GDP of $275.5 billion therefore has a GDP per capita of approximately $2,000. Its annual energy consumption was only 0.61 quad (0.64 EJ), making its Energy Intensity a mere 2,003 BTU (2,113 kJ) per dollar—a quarter of the US rate. Low standards of living and economic performance primarily accounts for such a meager number.
- Russia, with a population of 143 million and a GDP of $1.408 trillion therefore has a GDP per capita of approximately $9,800. Its annual energy consumption was 29.6 quad (31.2 EJ), for an Energy Intensity of 19,597 BTU (20,676 kJ) per dollar, or more than double that of the US. This may be due to harsh climatic conditions in most of Eastern Russia and the country's vast territorial space.
- Italy, with a population of 60 million and a GDP of $1.8 trillion therefore has a GDP per capita of approximately $31,000. Its annual Energy Intensity of 122.8 tons of oil equivalent makes it the most energy efficient country in the G8 and one of the most energy efficient in the industrial world, largely due to traditionally high energy prices which have resulted in more efficient company and consumer behaviours.

Of course, these numbers were produced with a mix of 2003 and 2004 figures, many of which are estimates. Actual mathematical models should use precise data of appropriate matching periods of study.

Several countries, like Sweden, Norway, France, and Canada, have made the transition to operating on low-carbon utilities. Norway and Canada have made the switch to hydro power; France relies on nuclear power. Since these countries have made the shift, they produce about a fifth of the carbon emissions in comparison to 13 other countries, like some including USA, Japan, and Italy.

Economic Energy Efficiency

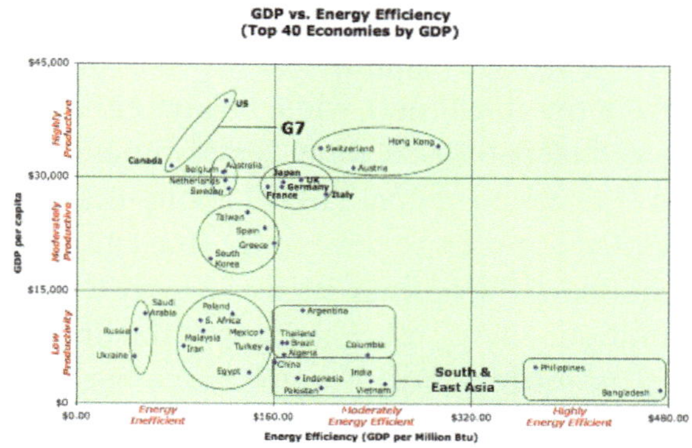

GDP per capita vs. 'Economic Energy Efficiency' plotted for the top 40 national economies.

An inverse way of looking at the issue would be an 'economic energy efficiency,' or economic rate of return on its consumption of energy: how many economic units of GDP are produced by the consumption of units of energy.

- Referring to the above examples, 1 million Btus consumed with an energy intensity of 8,553 produced $116.92 of GDP for the US. Whereas, each million Btus of energy consumed in Bangladesh with an Energy Intensity of 2,113 produced $473 of GDP, or over four times the effective US rate. Russia, on the other hand, produced only $48.37 GDP per 1 million Btu based on an energy intensity of 20,676. Thus, Bangladesh could be perceived as having nearly ten times the economic energy efficiency of Russia.

It is not directly causal that a high GDP per capita must have lower economic energy effi-ciencies.

Energy intensity can be used as a comparative measure between countries; whereas the change in energy consumption required to raise GDP in a specific country over time is described as its energy elasticity.

Energy Efficiency Gap

Energy efficiency gap refers to the improvement potential of energy efficiency or the difference between the cost-minimizing level of energy efficiency and the level of energy efficiency actually realized. It has attracted considerable attention among energy policy analysts, because its existence suggests that society has forgone cost-effective investments in energy efficiency, even though they could significantly reduce energy consumption at low cost. This term was first "coined" by Eric Hirst and Marilyn Brown in a paper entitled "Closing the Efficiency Gap: Barriers to the Efficient Use of Energy" in 1990.

Introduction

Energy efficiency refers to changes in equipment and behavior that result in increased energy services per unit of energy consumed, while behavioral changes that reduce energy use are often referred to as energy conservation. Energy intensity which measures energy consumption per Gross Domestic Product (GDP) is one indicator of energy efficiency. Many people have attempted to measure the energy efficiency gap, and their approaches differ based on the definitions of the optimal level of energy use. A popular theme is Hirst and Brown (1990)'s definition: energy efficiency gap is the unexploited economic potential for energy efficiency, in other words, it emphasizes the technically feasible energy efficiency measures that are cost-effective but are not being deployed. Many other studies have used this definition, such as International Energy Agency (2007) and Koopmans and Velde. Jaffe and Stavins (1994) identify five types of optimality and the corresponding definitions of the energy-efficiency gap: the economists' economic potential, the technologists' economic potential, hypothetical potential, the narrow social optimum and the true social optimum. In particular, economists' economic potential could be achieved by eliminating market failures in the energy efficiency technology market, while technologists' economic potential could be achieved by eliminating both market and non-market failures. Achieving the hypothetical potential would require the elimination of market failures in the whole energy market, for instance, having energy prices that reflect all externalities. The society can achieve the narrow social optimum by implementing all available cost-effective programs, and the true social optimum can be achieved if the environmental effects of energy generation and consumption is taken into consideration.

Barriers for Energy Efficiency Gap

Energy efficiency gaps exist because market failures exist. It is important to identify and understand those barriers in order to achieve desirable government policy interventions. According to Hirst and Brown (1990), various barriers that prevent the society from successfully closing energy efficiency gap can be divided into two categories: structural barriers and behavioral barriers. Structural barriers result from the actions of public and private organizations, and are usually beyond the control of the individual energy end user. Some examples are presented as following:

Distortion in fuel prices. The fuel prices that consumers pay do not reflect the social and environmental costs associated with fuel production, distribution and consumption. Consumers tend not to invest on energy efficiency technologies due to this distortion.

Uncertainty about future fuel prices. There have been great uncertainties with the prices for fuels, such as electricity and petroleum. More stringent environmental regulations and global warming concerns also increase the volatility of fuel prices. These uncertainties prevent consumers from making rational purchase decisions of new energy-using systems.

Limited access to capital. Consumers often face high up-front costs for energy-efficient systems. In addition, high discount rates are used to make tradeoffs between the initial capital investment and reduced operating costs, which also hinder the investments in energy-efficiency technologies.

Government fiscal and regulatory policies. Government policies tend to encourage energy consumption, rather than energy efficiency. For instance, government support has focused more on energy production, and the profit of electric utilities is a function of sales.

Codes and standards.The development of codes and standards often lag behind the development of technologies. It also takes a long time to adopt and modify standards, which becomes a barrier for energy efficiency technological innovation.

Supply infrastructure limitations. The deployment of energy efficiency technologies is highly restricted by factors such as geography, infrastructure and human resources.

Behavioral barriers are problems that characterize the end-user's decision-marking relating to energy consumption. Four examples are discussed below.

Attitudes toward energy efficiency.Public's awareness and attitudes toward energy efficiency could greatly affect their energy-related purchase and consumption behaviors.

Perceived risk of energy-efficiency investments. Consumers and businesses can be very risk averse in terms of investing in energy efficiency technologies. The uncertainties of fuel prices and high discount rate for operating costs have both made energy-efficiency investments even more "risky" for many decision makers.

Information gaps. There is often a lack of information on the performance of energy-efficient technologies. Consumers tend not to change their energy consumption behavior if little information is provided.

Misplaced incentives. The principal-agent problem and a lack of life-cycle thinking on costs and savings have imposed barriers for energy conservation.

Jaffe and Stavins (1994) categorize the barriers differently. They think that both market failures and non-market failures could account for the limited market success of the cost-effective energy-efficiency technologies. One important source of market failure is imperfect information, for instance the public good attributes of information and information asymmetry. Non-market failure may include the heterogeneity and inertia of consumers, and uncertainty about future energy prices and actual savings from energy efficiency investments.

Measures narrowing the Energy Efficiency Gap

Energy efficiency gap exists in various sectors, ranging from households, small businesses, corporations, and governments. Many policies and programs have been developed to overcome those barriers and close the energy efficiency gap. Subsidies and incentives for energy-efficient technologies. Insufficient capital investment could be overcome by more aggressive tax subsidies, loan guarantees, and low-interest government loans for energy efficient technologies.

Minimum building and equipment efficiency standards. Minimum building and equipment efficiency standards are cost-effective approaches to save energy. Effective implementation and upgrading of building energy efficiency standards could improve the energy integrity of new buildings, while equipment efficiency standards could help reduce energy consumption and pollution during the life-cycle of equipment.

Information programs. Research has proved that providing accurate and trustworthy information on energy use and energy efficiency choices could help narrow the gap. Three forms of information programs can be implemented to help producers and consumers make more informed and rational decisions. The first one is generic information applicable to all energy decision, such as forecasts of future energy pricies; the second type of program is to provide comparative information to facilitate technology and product choices, such as product rating and labeling systems; the third type of program is to offer specific recommendations for producers' and consumers' investment choices or behavior changes.

Government procurement programs for energy-efficient technologies. Government agencies could be required to procure energy-efficient products. This would help improve the energy efficiency of government sector, and the "learning by doing" impact would create early markets for energy-efficient technologies. Some real-world examples of those measures include the following: EU's energy consumption labeling scheme, U.S. DOE's building energy codes program, and EPA's and DOE's ENERGY STAR® voluntary labeling programs.

References

- "Assessment of Fuel Economy Technologies for Light-duty Vehicles". The National Academies Press. 2011. ISBN 978-0-309-15607-3. Retrieved 2016-09-18.
- Barney L. Capehart (2007). Encyclopedia of Energy Engineering and Technology, Volume 1. CRC Press. ISBN 0-8493-3653-8, ISBN 978-0-8493-3653-9.
- "Pathways to a Low-Carbon Economy: Version 2 of the Global Greenhouse Gas Abatement Cost Curve". McKinsey Global Institute: 7. 2009. Retrieved February 16, 2016.

- The Energy of the Future: Fourth "Energy Transition" Monitoring Report — Summary (PDF). Berlin, Germany: Federal Ministry for Economic Affairs and Energy (BMWi). November 2015. Retrieved 2016-06-09.
- Schlomann, Barbara; Eichhammer, Wolfgang (2012). Energy efficiency policies and measures in Germany (PDF). Karlsruhe, Germany: Fraunhofer Institute for Systems and Innovation Research ISI. Retrieved 2016-05-01.
- Agora Energiewende (2014). Benefits of energy efficiency on the German power sector : summary of key findings from a study conducted by Prognos AG and IAEW (PDF). Berlin, Germany: Agora Energiewende. Retrieved 2016-04-29.
- "National Action Plan on Energy Efficiency (NAPE): making more out of energy". Federal Ministry for Economic Affairs and Energy (BMWi). Retrieved 2016-06-07.
- Making more out of energy: National Action Plan on Energy Efficiency (PDF). Berlin, Germany: Federal Ministry for Economic Affairs and Energy (BMWi). December 2014. Retrieved 2016-06-07.
- "Gabriel: Efficiency First — discuss the Green Paper on Energy Efficiency with us!" (Press release). Berlin, Germany: Federal Ministry for Economic Affairs and Energy (BMWi). 12 August 2016. Retrieved 2016-09-06.
- Amelang, Sören (15 August 2016). "Lagging efficiency to get top priority in Germany's Energiewende". Clean Energy Wire (CLEW). Berlin, Germany. Retrieved 2016-09-06.
- Sekuła-Baranska, Sandra (24 May 2016). "New Act on Energy Efficiency passed in Poland". Noerr. Munich, Germany. Retrieved 2016-09-20.
- Huntington, Hillard (2011). EMF 25: Energy efficiency and climate change mitigation — Executive summary report (volume 1) (PDF). Stanford, CA, USA: Energy Modeling Forum. Retrieved 2016-05-10.

4

Key Concepts of Energy Conservation

New technologies have increased the efficiency of resource use and the benefits caused by these technologies are known as the rebound effect. Rebound effect, negawatt power, Earth hour, energy hierarchy and one-watt initiative are some of the key movements of energy conservation. The following chapter unfolds its crucial aspects in a critical yet systematic manner.

Rebound Effect (Conservation)

In conservation and energy economics, the rebound effect (or take-back effect, RE) is the reduction in expected gains from new technologies that increase the efficiency of resource use, because of behavioral or other systemic responses. These responses usually tend to offset the beneficial effects of the new technology or other measures taken. While the literature on the rebound effect generally focuses on the effect of technological improvements on energy consumption, the theory can also be applied to the use of any natural resource or other input, such as labor. The rebound effect is generally expressed as a ratio of the lost benefit compared to the expected environmental benefit when holding consumption constant.

For instance, if a 5% improvement in vehicle fuel efficiency results in only a 2% drop in fuel use, there is a 60% rebound effect (since $^{(5-2)}/_{5} = 60\%$). The 'missing' 3% might have been consumed by driving faster or further than before.

The existence of the rebound effect is uncontroversial. However, debate continues as to the magnitude and impact of the effect in real world situations. Depending on the magnitude of the rebound effect, there are five different rebound effect types:

1. Super conservation (RE < 0): the actual resource savings are higher than expected savings – the rebound effect is negative. This occurs if the increase in efficiency reduces costs.

2. Zero rebound (RE = 0): The actual resource savings are equal to expected savings – the rebound effect is zero.

3. Partial rebound (0 < RE < 1): The actual resource savings are less than expected savings – the rebound effect is between 0% and 100%. This is sometimes known as 'take-back', and is the most common result of empirical studies on individual markets.

4. Full rebound (RE = 1): The actual resource savings are equal to the increase in usage – the rebound effect is at 100%.

5. Backfire (RE > 1): The actual resource savings are negative because usage increased beyond potential savings – the rebound effect is higher than 100%. This situation is commonly known as the Jevons paradox.

In order to avoid the rebound effect, environmental economists have suggested that any cost savings from efficiency gains be taxed in order to keep the cost of use the same.

History

The rebound effect was first described by William Stanley Jevons in his 1865 book *The Coal Question*, where he observed that the invention in Britain of a more efficient steam engine meant that the use of coal became economically viable for many new uses. This ultimately led to increased coal demand and much increased coal consumption, even as the amount of coal required for any particular use fell. According to Jevons, "It is a confusion of ideas to suppose that the economical use of fuel is equivalent to diminished consumption. The very contrary is the truth."

However, most contemporary authors credit Daniel Khazzoom for the re-emergence of the rebound effect in the research literature. Although Khazzoom did not use the term, he raised the idea that there is a less than one-to-one correlation between gains in energy efficiency and reductions in energy use, because of a change in the 'price content' of energy in the provision of the final consumer product. His study was based on energy efficiency gains in home appliances, but the principle applies throughout the economy. A commonly studied example is that of a more fuel-efficient car. As each kilometre of travel becomes cheaper, there will be an increase in driving speed and/or kilometres driven, as long as the price elasticity of demand for car travel is not zero. Other examples might include the growth in garden lighting after the introduction of energy-saving compact fluorescent lamps or the increasing size of houses driven partly by higher fuel efficiency in home heating technologies. If the rebound effect is larger than 100%, all gains from the increased fuel efficiency would be wiped out by increases in demand (the Jevons paradox).

Khazzoom's thesis was criticized heavily by Michael Grubb and Amory Lovins who dismissed any disconnection between energy efficiency improvements in an individual market, and an economy-wide reduction in energy consumption. Developing Khazzoom's idea further, and prompting heated debate in the Energy Policy journal at that time, Len Brookes wrote of the fallacies in the energy-efficiency solution to greenhouse gas emissions. His analysis showed that any economically justified improvements in energy efficiency would in fact stimulate economic growth and increase total energy use. For improvements in energy efficiency to contribute to a reduction in economy-wide energy consumption, the improvement must come at a greater economic cost.

Commenting in regard to energy efficiency advocates, he concludes that, "the present high profile of the topic seems to owe more to the current tide of green fervor than to sober consideration of the facts, and the validity and cost of solutions."

Khazzoom-Brookes Postulate

In 1992, economist Harry Saunders coined the term "Khazzoom-Brookes postulate" to describe the idea that energy efficiency gains paradoxically result in increases in energy use (the modern day equivalent of the Jevons paradox). He modeled energy efficiency gains using a variety of neo-classical growth models, and showed that the postulate is true over a wide range of assumptions. In the conclusion of his paper, Saunders stated that:

In the absence of efficiency gains, energy use will grow in lock step with economic growth (energy intensity will stay fixed) when energy prices are fixed. ... Energy efficiency gains can increase energy consumption by two means: by making energy appear effectively cheaper than other inputs; and by increasing economic growth, which pulls up energy use. ... These results, while by no means proving the Khazzoom-Brookes postulate, call for prudent energy analysts and policy makers to pause a long moment before dismissing it.

This work provided a theoretical grounding for empirical studies and played an important role in framing the problem of the rebound effect. It also reinforced an emerging ideological divide between energy economists on the extent of the yet to be named effect. The two tightly held positions are:

- Technological improvements in energy efficiency enable economic growth that was otherwise impossible without the improvement; as such, energy efficiency improvements will usually back-fire in the long term.
- Technological improvements in energy efficiency may result in a small take-back. However, even in the long term, energy efficiency improvements usually result in large overall energy savings.

Even though many studies have been undertaken in this area, neither position has yet claimed a consensus view in the academic literature. Recent studies have demonstrated that direct rebound effects are significant (about 30% for energy), but that there is not enough information about indirect effects to know whether or how often back-fire occurs. Economists tend to the first position, but most governments, businesses, and environmental groups adhere to the second. Governments and environmental groups often advocate further research into fuel efficiency and radical increases in the efficient use of energy as the primary means for reducing energy use and reducing greenhouse gas emissions (to alleviate the impacts of climate change). However, if the first position more accurately reflects economic reality, current efforts to invent fuel-efficient technologies may not much reduce energy use, and may in fact paradoxically increase oil and coal consumption, and greenhouse gas emissions, over the long run.

Types of Effects

The full rebound effect can be distinguished into three different economic reactions to technological changes:

1. Direct rebound effect: An increase in consumption of a good is caused by the lower cost of use. This is caused by the substitution effect.

2. Indirect rebound effect: The lower cost of a service enables increased household consumption of other goods and services. For example, the savings from a more efficient cooling system may be put into another luxury good. This is caused by the income effect.

3. Economy wide effect: The fall in service cost reduces the price of other goods, creates new production possibilities and increases economic growth.

In the example of improved vehicle fuel efficiency, the direct effect would be the increased fuel use from more driving as driving becomes cheaper. The indirect effect would incorporate the increased consumption of other goods enabled by household cost savings from increased fuel efficiency. Since consumption of other goods increase, the embodied fuel used in the production of those goods would increase as well. Finally, the economy wide effect would include the long-term effect of the increase in vehicle fuel efficiency on production and consumption possibilities throughout the economy, including any effects on economic growth rates.

Direct and Indirect Effects

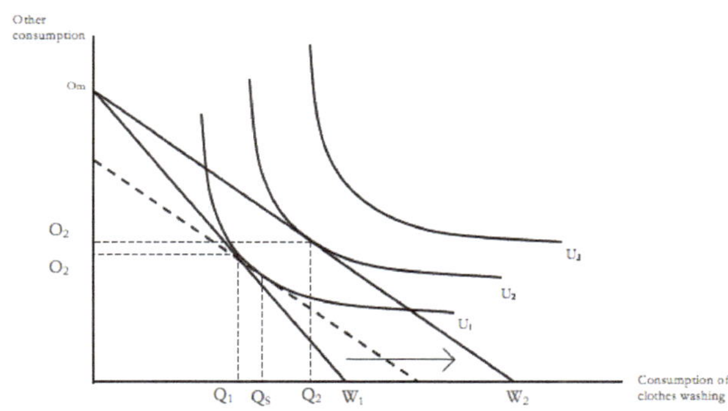

Direct and Indirect Effects

For cost reducing resource efficiency, distinguishing between direct and indirect effects is shown in Figure 1 below. The horizontal axis shows units of consumption of the targets good (which could be for example clothes washing, and measured in terms of kilograms of clean clothes) with consumption of all other goods and services on the vertical axis. An economical technology change that enables each unit of washing to be produced with less electricity results in a reduction of the price per unit of washing. This shifts the household budget line rightwards. The result is a substitution effect because of the decreased relative price, but also an income effect due to the increased real income. The substitution effect increases consumption of washing from Q1 to QS, and the income effect from QS to Q2. The total increase in consumption of washing from Q1 to Q2 and the resulting increase in electricity consumption is the direct effect. The indirect effect comprises the increase in other consumption, from O1 to O2. The scale of each of these effects depends on the elasticity of demand for each of the goods, and the embodied resource or externality associated with each good. Indirect effects are difficult to measure empirically. In the manufacturing sector, it has been estimated that there is about a 24% rebound effect due to increases in fuel efficiency. A parallel effect will happen for cost saving efficient technologies for producers, where output and substitution effects will occur.

The rebound effect can increase the difficulty of projecting the reduction in greenhouse emissions from an improvement in energy efficiency. Estimation of the scale of direct effects on residential electricity, heating and motor fuel consumption has been common motivation for research of rebound effects. Evaluation and econometric methods are the two approaches generally employed in estimating the size of this effect. Evaluation methods rely on quasi-experimental studies and measure the before and after changes to energy consumption from the implementation of energy efficient technology, while econometric methods utilize elasticity estimates to forecast the likely effects from changes in the effective price of energy services.

Research has found that in developed countries, the direct rebound effect is usually small to moderate, ranging from roughly 5% to 40% in residential space heating and cooling. Some of the direct rebound effect can be attributed to consumers who were previously unable to use a service. However, the rebound effect may be more significant in the context of the undeveloped markets in developing economies.

Indirect Effects from Conservation

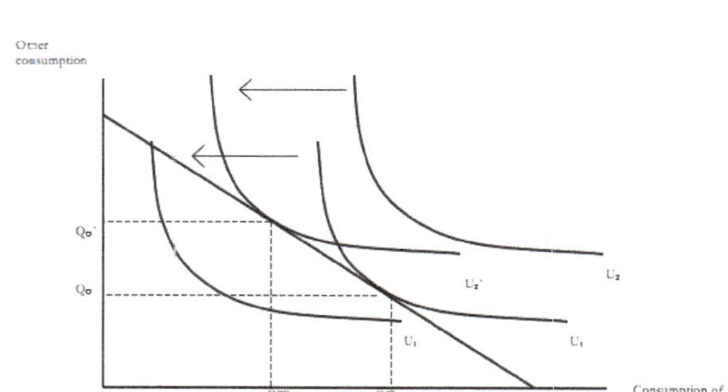

Figure2: change in preferences of a household revealing indirect effects from conservation

For conservation measures, indirect effects closely approximate the total economy wide effect. Conservation measures constitute a change in consumption patterns away from particular targeted goods towards other goods. Figure 2 shows that a change in preference of a household results in a new consumption pattern that has less of the target good (QT to QT`), and more of all other goods (QO to QO`). The resource consumption or externalities embodied in this other consumption is the indirect effect.

Although a persuasive view has prevailed that indirect effects with respect to energy and greenhouse emissions should be very small due to energy directly comprising only a small component of household expenditure, this view is gradually being eroded. Many recent studies based on life-cycle analysis show the energy consumed indirectly by households is often higher than consumed directly through electricity, gas, and motor fuel, and is a growing proportion. This is evident in the results of recent studies that indicate indirect effects from household conservation can range from 10% to 200% depending on the scenario, with higher indirect rebounds from diet changes aiming to reduce food miles.

Economy Wide Effects

Even if the direct and indirect rebound effects add up to less than 100%, technological improvements that increase efficiency may still result in economy wide effects that results in increased resource use for the economy as a whole. In particular, this would happen if resource efficiency enables an expansion of production in the economy, and an increase in the rate of economic growth. For example, for the case of energy use, more efficient technology is equivalent to a lower price for energy resources. It is well known that changes in energy costs have a large impact on economic growth rates. In the 1970s sharp increases in petroleum prices led to stagflation (recession and inflation) in the developed countries, whereas in the 1990s lower petroleum prices contributed to higher economic growth. An improvement in energy efficiency has the same effect as lower fuel prices, and leads to faster economic growth. Economists generally believe that especially for the case of energy use, more efficient technologies will lead to increased use, because of this growth effect.

To model the scale of this effect, economists use computational general equilibrium (CGE) models. While CGE methodology is by no means perfect, results indicate that economy wide rebound effects are likely to be very high, with estimates above 100% rather common. One simple CGE model has been made available online for use by economists.

Income Level Variation

Research has shown that the direct rebound effects for energy services is lower at high income levels, due to less price sensitivity. Studies have found that own-price elasticity of gas consumption by UK households was two times greater for households in the lowest income decile when compared to the highest decile. Studies have also observed higher rebounds in low-income houses for improvements in heating technology. Evaluation methods have also been used to assess the scale of rebound effects from efficient heating installations in lower income homes in the United Kingdom. This research found that direct effects are close to 100% in many cases. High income households in developed countries are likely to set the temperature at the optimum comfort level, regardless of the cost – therefore any cost reduction does not result in increased heating, for it was already optimal. But low-income households are more price sensitive, and have made thermal sacrifices due to the cost of heating. In this case, a high direct rebound is likely. This analogy can be extended to most household energy consumption.

The size of the rebound effect is likely to be higher in developing countries according to macro-level assessments and case studies. One case study was undertaken in rural India to evaluate the impact of an alternative energy scheme. Households were given solar powered lighting in an attempt to reduce the use of kerosene for lighting to zero except for seasons with insufficient sunshine. The scheme was also designed to encourage a future willingness to pay for efficient lighting. The results were surprising, with high direct rebounds between 50 and 80%, and total direct and indirect rebound above 100%. Because the new lighting source was essentially zero cost, operating hours for lighting went up from an average of 2 to 6 per day, with new lighting consisting of a combination of both the no-cost solar lamps and also kerosene lamps. Also, more cooking was undertaken which enabled an increased trade of food with neighboring villages.

Rebounds With Respect to Time

The individual opportunity of cost is an often overlooked cause of rebound effect. Time is being substituted with an increase in demand for a service. Research articles often examine increasingly convenient and more rapid modes of transportation to determine the rebound effect in energy demand. Because time cost forms a major part of the total cost of commuter transport, rapid modes will reduce real costs, but will also encourage longer commuting distances which will in turn increase energy consumption. While important, it is almost impossible to estimate empirically the scale of such effects due to the subjective nature of the value of time. Time saved can either be used towards additional work or leisure which may have differing degrees of rebound effect. Labor time saved at work due to the increased labour productivity is likely to be spent on further labor time at higher productive rates. For leisure time saving, this may simply encourage people to diversify their leisure interests to fill their generally fixed period of leisure time.

Suggested Solutions

In order to ensure that efficiency enhancing technological improvements actually reduce fuel use, the ecological economists Mathis Wackernagel and William Rees have suggested that any cost savings from efficiency gains be "taxed away or otherwise removed from further economic circulation. Preferably they should be captured for reinvestment in natural capital rehabilitation." This can be achieved through, for example, the imposition of a green tax, a cap and trade program, higher fuel taxes or the proposed "restore" approach. Policies can also directly address projected yearly consumption of energy rather than device efficiency, especially for systems where the use can be accurately projected, such as street lighting.

Negawatt Power

A "negawatt revolution" would involve the rapid deployment of electricity-saving technologies, such as compact fluorescent lamps.

Negawatt power is a theoretical unit of power representing an amount of energy (measured in watts) saved. The energy saved is a direct result of energy conservation or increased energy efficiency. The term was coined by the chief scientist of the Rocky Mountain Institute and environmentalist Amory Lovins in 1985, within the article, "Saving Gigabucks with Negawatts," where he argued that utility customers don't want kilowatt-hours of electricity; they want energy services such as hot showers, cold beer, lit rooms, and spinning shafts, which can come more cheaply if electricity is used more efficiently. Lovins felt an international behavioral change was necessary in order to decrease countries' dependence on excessive amounts of energy. The concept of a negawatt could influence a behavioral change in consumers by encouraging them to think about the energy that they spend.

A negawatt market can be thought of as a secondary market, in which electricity is allocated from one consumer to another consumer within the energy market. In this market, negawatts could be treated as a commodity. Commodities have the ability to be traded across time and space, which would allow negawatts to be incorporated in the international trading system. Roughly 10% of all U.S. electrical generating capacity is in place to meet the last 1% of demand and there is where the immediate efficiency opportunity exists.

On March 15, 2011, the Federal Energy Regulatory Commission (FERC), the agency that regulates the U.S. electrical grid, approved a rule establishing the approach to compensation for demand response resources intended to benefit customers and help improve the operation and competitiveness of organized wholesale energy markets.

This means that negawatts produced by reducing electrical use can demand the same market prices as real megawatts of generated electricity.

The incentives for a negawatt market include receiving money, reduction of national energy dependency, and the local electricity deregulation within certain nations or states. As for the cost incentive, those who produce negawatts or simply conserve energy can earn money by selling the saved energy. The negawatt market could help nations or states obtain a deregulated electricity system by creating another market to purchase electricity from.

The negawatt market also has two main drawbacks. Currently, there is no way to precisely measure the amount of energy saved in negawatts, and electricity providers may not want customers to use less energy due to the loss of profit.

An assortment of LED lamps commercially available as of 2010 as replacements for screw-in bulbs, including floodlight fixtures (left), reading light (center), household lamps (center right and bottom), and low-power accent light (right) applications

Definition

Negawatt power is a theoretical unit of power representing an amount of energy (measured in Watts) saved. Energy is saved by either increased efficiency or reduced energy consumption; the conserved energy is a "negawatt". The concept of a negawatt is simply a measure of power that is not used. Negawatts are a form of encouragement to motivate consumers to conserve energy. Amory Lovins considers the concept of conservation "a change in behavior based on the attitude 'Do Less to Use Less.'" He makes a distinction between conservation and efficiency by defining efficiency as "the application of technologies and best practices to eliminate waste based on the attitude, 'Do the same or more with less.'"

Cost for negawatt power can be calculated using cost-effectiveness analysis or CEA. For energy efficiency investments a CEA calculation produces the value of saved energy or *negawatts* in \$/kWh. Such a valuation allows comparing the price of negawatts with price of energy such as electricity from the grid or the cheapest renewable alternative.

Specifically, Tuominen et al. have suggested using the dynamic generation cost type of CEA for energy efficiency investments as it includes the best accounting for the time value among the various CEA methods available.

Origin and Development

A negawatt is a unit in watts of energy saved. It is basically the opposite of a watt. The term negawatt is derived from megawatt and was created by Amory Lovins. Lovins saw a typo – "negawatt" instead of "megawatt" – in a Colorado Public Utilities Commission report in 1989. He adopted the term to describe electricity that was not created using energy efficiency and conservation. He was concerned with the large inefficiencies of energy use and came up with ways to remedy the problem. Lovins advocated for more efficient light bulbs and reflective metals that increase the intensity of light produced. He wanted consumers to use the energy produced in a smarter way by "wringing more work from the electricity we already have." Lovins felt an international behavioral change was necessary in order to decrease countries' dependence on excessive amounts of energy. The concept of a negawatt could influence a behavioral change in consumers by encouraging them to think about the energy that they spend.

According to Lovins, energy efficiency represents a profitable global market and American companies have at their disposal the technical innovations to lead the way. Not only should they "upgrade their plants and office buildings, but they should encourage the formation of negawatt markets". Lovins sees negawatt markets as a win-win solution to many environmental problems. Because it is "now generally cheaper to save fuel than to burn it, global warming, acid rain, and urban smog can be reduced not at a cost but at a profit".

Lovins explains that many companies are already enjoying the financial and other rewards that come from saving electricity. Yet progress in converting to electricity saving technologies has been slowed by the indifference or outright opposition of some utilities. A second obstacle to efficiency is that many electricity-using devices are purchased by people who won't be paying their running costs and thus have little incentive to consider efficiency. Lovins also believes that many customers "don't know what the best efficiency buys are, where to get them, or how to shop for them".

In 2003 in France under the guide of Thierry Salomon 23 scientists wrote "Le manifeste Négawatt." Megawatt and negawatt seem to be reminiscent to the larger concept of ecological footprint, handprint handprint, and by following this line of thought toward compatibility and comparability a second frame of concept seems appropriate: mindprint – the impact in another frame or setting where units or numbers can not be compared.

Market

Amory Lovins has advocated a "negawatt revolution", arguing that utility customers don't want kilowatt-hours of electricity; they want energy services such as hot showers, cold beer, lit rooms, and spinning shafts, which can come more cheaply if electricity is used more efficiently. Lovins, defines the negawatt market as a way to reduce the gap between the cost of making and saving electricity.

The negawatt market can be thought of as a secondary market where electricity is allocated from areas of less use to areas of greater use. This would be a secondary market, due to the fact that it would reallocate electricity from one consumer to another within the already existing energy market. Some feel that to establish a viable market, legislation and cooperation between primary producers, distributors, traders and consumers, may be required. This proposal would encourage the market to have legislative regulations, while still allowing the market to work within itself to set prices and allocate resources.

A negawatt market would allow "demand side resources" to participate in wholesale energy markets. These markets are commonly referred to as a demand response. Demand response can be defined as "enrolling large users of energy in programs to lower their usage in return for compensation, which helps take pressure off the grid" This market would help take pressure off the grid because electricity could "be treated as a commodity just like copper or sowbellies," and therefore traded to areas that need it more than others. As any commodity, negawatts would have to be "tradable across time and space" to be an effective market. Being able to trade negawatts across time and space would allow for an international trading system. To create a market for negawatts, energy companies will need to put more focus toward energy efficiency. This shift in focus would require a new "business structure that will thrive in the 'negawatt market'", which has not yet been developed. Market possibilities are being implemented internationally, implying that one day an official negawatt market will become a reality.

Implementation

Government Implementation

Negawatt power is being implemented in many states in the U.S. and is emerging as an international strategy to reduce energy consumption. "Test negawatt auctions began in 1999 in Connecticut and Georgia and more than a dozen utility exchanges were in existence" in 2000. In an effort to move toward energy efficiency, New York has created programs "supported through Energy $mart, which is run by the New York State Energy Research and Development Authority (NYSERDA), with money from a small surcharge on utility bills." Negawatt power is implemented in California as well as Texas. "Some Texas congressmen and energy companies are trying to help California avert blackouts and utility price shocks this summer with [...] 'negawatts'.

On January 1, 2009, the states of South Australia and Victoria (Australia) became the first in Australia to offer "householders energy efficiency incentives programs delivered via local electricity retailers."

The UK is looking to balance the demand of electricity by proposing a bill in which "the government may pay major users for each "negawatt" of capacity they switch off" during high use. Payment may also be issued for supplying the grid.

Private Implementation

The negawatt market is being used by governments and companies. Aluminum manufacturers in the Pacific Northwest shut down their power plants and sold the unused energy because selling the negawatts was more profitable for the company than selling the aluminum product. This was possible because "The smelters hold power contracts with the federal Bonneville Power Administration that contain clauses allowing them to market the electricity."

The Associated Electric company in rural Missouri is implementing the usage and spreading the knowledge of negawatts by performing energy audits at their customer's homes to show them where they could be saving electricity. Rebates are being given to help cutomers pay for more energy-efficient, Energy Star appliances. Keith Hartner, the CEO of Associated Electric Cooperative Inc., feels that negawatts are generating savings for their customers and for the company as well: "The goal of this program is to save money not only at the generator but also at the meter for the members."

Individual households can practice negawatts through using energy-efficient lighting and Energy Star appliances as well as simply reducing standby power. The resulting savings sometimes offset the investment in new high efficient light bulbs and appliances. These efficiencies can offset a portion of the cost of a residential photovoltaic system. Negawatts reduces the overall size of the photovoltaic system required for the household, because they are consuming less electricity. This results in a faster payback period for the photovoltaic system. The City of San Diego has created a negawatts initiative called "Reduce then Produce" to promote this idea.

Advantages

Cost

The most noteworthy advantage in creating a negawatt market is the cost incentive. As many will say, "The cheapest watt is the one that's never created." In this market, the consumers who increase their home efficiency, or decrease their energy consumption, can earn money by selling the saved electricity. This is similar to an emissions trading or a cap-and-trade system, in which the energy that is not used can be bought from the consumers who saved the energy and sold to those who need to purchase the extra energy. Providers of electricity can encourage consumers to sell back their unused energy, or negawatts, especially during peak hours. A major question that electric companies need to ask themselves is whether it is less expensive to pay consumers to reduce consumption for a few instances a year or to build and maintain a side-supply resource that would only be used a few times a year. Many argue that the "cost of foregone consumption is less than the cost of increasing the supply of electricity."

If a consumer conserves a substantial amount of energy then there can even be an allowance for a tax deduction. According to the Negawatt Power Solutions Group, a "building that achieved a 50% energy cost reduction may be eligible for tax deduction up to $1.80 per square foot." Negawatts can help alleviate some of the costs of constructing new, efficient buildings. "The negawatt revolution now provides a way to cut construction costs, capture big returns on capital in renovations, [and] dramatically cut operating expenses." Existing buildings can be made more efficient by renovating the insulation to cut back on electricity used for heating, installing more efficient light fixtures, and an upgraded HVAC design. Renovating a building to be more energy efficient will cost less than creating entirely new, more efficient buildings.

Reduction of National Energy Dependency

The reduction of the amount of energy that a region emits can slowly separate a nation from a high energy consumption of oil. The desire to become a less energy-dependent country may promote behavioral changes and a different societal outlook on energy usage. These potential societal perspective changes could lead to an overall more sustainable country. The reduced consumption of energy would also produce less greenhouse gases, which could have positive outcomes on the economy, political parties, and interest groups, such as environmentalists. According to Lovins, improvements in energy efficiency and conservation, due to a change in behavioral attitudes, have a huge potential to reduce a country's "long-term energy needs," such as the United States.

Local Deregulation

Some conservatives claim that the negawatt market could help nations or states have a deregulated electricity system. This would allow a nation or a state to experiment with "electricity deregulation," in which "demand reductions could be purchased with a minimum of disruption to businesses, workers and the economy." In the United States, for instance, the negawatt market could assist California with rolling blackouts by making more power available from consumers who choose to conserve energy or increase their negawatts. California could achieve the goal of deregulation by allowing a deficit area to "purchase an emergency supply from anywhere within with West" in which "the ultimate purpose of deregulation was to allow competition in the electricity market and consumer choice of electricity providers." Negawatt power would allow the consumers in a country's economy to decide how the energy will be distributed: essentially benefiting regions that hope to have a deregulated electricity system.

Drawbacks

Difficulty in Creating A Negawatt Market

Currently, there is no way to precisely measure the amount of energy saved in negawatts; it can only be theoretically determined based on the consumer's history of energy use. Visualizing has a very important role in "enabling residents to understand and manage their energy use," which serves as a form of encouragement for consumers to conserve energy. Without the visualization of the energy use, it is difficult to conceptualize negawatts because the consumer cannot see a precise value of the amount of saved energy. Smart meters are becoming a more developed technology to measure energy usage, but "consumers are calling on state regulators to move cautiously on smart meters, citing complaints in some states that the meters are raising electric bills rather than lowering them."

Some municipally owned utilities and cooperatives argue that negawatt power "lets consumers treat electricity as a property right rather than a service [...giving them] legal entitlement to power [that they] don't consume." This would indicate that consumers would treat electricity as a property, not a service. Some people, including the senior vice president Joe Nipper from the American Public Power Association, oppose the idea that people would receive money for power that they did not even spend.

Electricity price caps may also need to be implemented in order for the emerging negawatts market to function correctly. To some, government limitations on markets are unfavorable. There is a current view that negawatts are worth pursuing, but that they are unlikely to satisfy the world's thirst for energy to the extent their advocates assume.

Expense of Efficiency

Saving energy by the negawatt and creating a negawatt market can present several drawbacks for manufacturers and electricity providers. Manufacturers are less inclined to make energy-efficient devices which meet a specific standard, such as Energy star's standard, because of increased time and cost, while receiving minimal profit. Overall, electricity providers may not want customers to use less energy due to the loss of profit. Some even argue that producing energy-efficient products, such as light bulbs, actually simulate more demand, "resulting in more energy being purchased for conversion into light" of the same strength.

Customers may also be less inclined to buy products that are more energy efficient due to the increase in cost and time spent. "Even when the information is known and, despite the overall long-term cost-saving potential, the price of energy is too low...for individuals to justify the initial cost of energy efficiency measures." Not only are energy efficient devices more expensive, but "consumers are poorly informed about the savings on offer. Even when they can do the sums, the transaction costs are high: it is a time-consuming chore for someone to identify the best energy-saving equipment, buy it and get it installed."

The technology used to measure the amount of energy that a consumer uses and saves, known as smart meters, grid systems, or energy dashboards, require time for the consumer to understand. Some argue that people need to have access to "simple yet effective information systems to help users understand their energy without having to become technology experts." The current visual energy measuring devices could be made simpler and less expensive, which may encourage customers to save more energy, thus increasing their negawatts.

The capital cost of energy-efficiency is erroneously argued by some to exceed the energy savings (energy required for raw materials, manufacture and transport to point of sale versus running costs). For instance they argue that the capital cost of buying a new energy efficient car will not be recouped unless the car runs for more than ten years. However increased energy efficiency leads to less usage and longer life.

Earth Hour

Earth Hour's logo

Earth Hour is a worldwide movement for the planet organized by the World Wide Fund for Nature (WWF). The event is held worldwide annually encouraging individuals, communities, households and businesses to turn off their non-essential lights for one hour, from 8:30 to 9:30 p.m. towards the end of March, as a symbol for their commitment to the planet. It was famously started as a lights-off event in Sydney, Australia, in 2007. Since then, it has grown to engage more than 7000 cities and towns worldwide. Today, Earth Hour engages a massive mainstream community on a broad range of environmental issues. The one-hour event continues to remain the key driver of the now larger movements.

Earth Hour 2016 was on Saturday, March 19, from 8:30 p.m. to 9:30 p.m. during participants' local time.

History

Conception and start: 2004–2007

In 2004, confronted with scientific findings, WWF Australia met with advertising agency Leo Burnett Sydney to "discuss ideas for engaging Australians on the issue of climate change". The idea of a large scale switch off was coined and developed in 2006, originally under the working title "The Big Flick". WWF Australia presented their concept to Fairfax Media who, along with Sydney Lord Mayor Clover Moore, agreed to back the event. The 2007 Earth Hour was held on March 31 in Sydney, Australia at 7:30 pm, local time.

In October 2007 San Francisco ran its own "Lights Out" program inspired by the Sydney Earth Hour. After their successful event in October, the organisers decided to rally behind the Earth Hour being planned for March 2008.

2008

Overview of Sydney during Earth Hour 2008

Earth Hour 2008 was held internationally on March 29, 2008 from 8 p.m. to 9 p.m. local time, marking the first anniversary of the event. With 35 countries around the world participating as official flagship cities and over 400 cities also supporting, Earth Hour 2008 was celebrated on all seven continents. Landmarks all around the world turned off their non-essential lighting for Earth Hour, including the Sydney Opera House (Sydney, Australia), Empire State Building (New York City), Sears Tower (now Willis Tower, Chicago, US), National Monument (Jakarta, Indonesia), Golden Gate Bridge (San Francisco, US), Bank of America Plaza (Atlanta, US), Space Needle (Seattle, US), Table Mountain (Cape Town, South Africa), the Colosseum (Rome, Italy), Azrieli Center (Tel Aviv, Israel), Royal Castle (Stockholm, Sweden), the CN Tower (Toronto, Canada), SM Mall of Asia, SM Science Discovery Center (Manila, Philippines), Suva (Fiji), Nidaros Cathedral (Trondheim, Norway), Petronas Towers (Kuala Lumpur, Malaysia), KL Tower (Kuala Lumpur, Malaysia), Wat Arun Temple (Bangkok, Thailand), The London City Hall (London, England) and the Royal Liver Building (Liverpool, United Kingdom),

The official website for the event, earthhour.org, received over 6.7 million unique visitors in the week leading up to Earth Hour. Other websites took part in the event, with Google's homepage going "dark" on the day Earth Hour took place.

According to a Zogby International online survey, 36 million Americans—approximately 16 percent of the United States adult population—participated in Earth Hour 2008. The survey also showed there was a 4 percentage point increase in the level of interest in environmental issues such as climate change and pollution directly after the event (73 percent pre-event versus 77 percent post-event).

2008 participants

Earth Hour 2008 included 26 major cities worldwide and 300 smaller towns shutting off their lights.

Scheduling

Sydney Harbour Bridge and Sydney Opera House were darkened during Earth Hour 2007.

Since Earth Hour for 2008 was on a Saturday, many high schools in the Greater Toronto Area participated by turning off half the lights in classrooms during the last hour of school on Friday, March 28, 2008. Although the tagline of Earth Hour 2008 was officially, "See the difference you can make", the official radio advertisement ended with the tagline, "Dark city, bright idea."

Azrieli Center in Tel Aviv darkened for Earth Hour 2010.

Tel Aviv scheduled their Earth Hour for Thursday March 27, 2008 to avoid conflict with Sabbath. Dublin moved their Earth Hour to between 9 and 10 p.m. due to their northern geographical location.

Reduced Energy Consumption

Colosseum darkened for Earth Hour 2008

Auditorio de Tenerife darkened for Earth Hour

According to WWF Thailand, Bangkok decreased electricity usage by 73.34 megawatts, which, over one hour, is equivalent to 41.6 tonnes of carbon dioxide. The *Bangkok Post* gave different figures of 165 megawatt-hours and 102 tonnes of carbon dioxide. This was noted to be significantly less than a similar campaign initiated by Bangkok's City Hall the previous year in May, when 530 megawatt-hours were saved and 143 tonnes of carbon dioxide emission were cut.

Philippine Electricity Market Corp. noted that power consumption dropped by about 78.63 megawatts in Metro Manila, and up to 102.2 megawatts on Luzon. The maximum demand drop of around 39 MW was experienced at 8:14 p.m. in Metro Manila and of around 116 MW at 8:34 p.m. in the Luzon grid.

Ontario used approximately 900 megawatt-hours less electrical energy during Earth Hour. At one point, Toronto, Ontario saw an 8.7% reduction in consumption as compared to a typical March Saturday night.

Ireland, as a whole, had a reduction in electricity use of about 1.5% for the evening. In the three-hour period between 6:30 p.m. and 9:30 pm, there was a reduction of 50 megawatts, saving 150 megawatt-hours, or approximately 60 tonnes of carbon dioxide.

Golden Gate Bridge and Marin Headlands public open space in the background, before (inset) and during Earth Hour 2008

In Dubai, where external lighting on several major city landmarks was turned off and street lighting in selected areas was dimmed by 50%, the Electricity and Water Authority reported savings of 100 megawatt-hours of electricity. This represented a 2.4% reduction in demand compared to before the hour began.

The Sky Tower in Auckland, New Zealand, switched off its usual floodlighting during the Earth Hour, and re-lit afterwards. (the red lights in the middle image are aircraft warning lights)

The best result was from Christchurch, New Zealand, with the city reporting a drop of 13% in electricity demand. However, national grid operator Transpower reported that New Zealand's power consumption during Earth Hour was 335 megawatts, higher than the 328 megawatt average of the previous two Saturdays. Melbourne, Australia reduced demand by 10.1%. Sydney, being the city that participated in both the 2007 and 2008 Earth Hours, cut electricity consumption by 8.4%. This is less than the previous year's 10.2%; however, Earth Hour executive director Andy Ridley made the claim that after factoring margin of error, the participation in this city was the same.

The worst result was from Calgary, Canada. The city's power consumption actually went up 3.6% at the hour's peak electricity demand. Calgary's weather plays a large role in power consumption, and the city experienced weather 12 °C (around 22 °F) colder than the previous Saturday's recorded temperature in the inaugural year. Enmax, the city's power supplier, has confirmed that in all subsequent years, Calgarians have not supported the Earth Hour initiative, noting that power consumption changed only marginally during the hour in 2010 and 2011 (1% or less) and in 2012 and 2013 showed no appreciable change in power usage at all.

Celebrations Around the World

- The Danish royal palaces, Amalienborg Palace and Gråsten Palace, went dark at the Queen's command.
- Nelly Furtado held a free concert at Nathan Phillips Square in Downtown Toronto to celebrate Earth Hour.
- In Toronto, Ontario, York University's student-run *Environmental Outreach Team* ran an afternoon Earth Hour information session, and the York University Observatory offered an extra public viewing session.
- Stargazing activities were held in Toronto's Ontario Science Centre and Richmond Hill's David Dunlap Observatory.
- Astronomy Ireland set up high-powered telescopes in Dublin's Phoenix Park to allow people to take advantage of the night sky, normally swamped by bright city lights.
- In Tel Aviv, Israel, a free concert by Knesiyat Hasekhel was held at Rabin Square. Power needed for the concert was generated by a group of cyclists pushing pedal generators. The rest of the power was supplied by generators burning used falafel oil for power.
- In Atlanta, the CEO of WWF US, Carter Roberts, and the Mayor of Atlanta, Shirley Franklin, flipped a giant switch on live TV, symbolically starting the wave of lights going out on the buildings around the city.
- In San Francisco, a public event hosted by WWF US was attended by Mayor Gavin Newsom, Gold medal figure skater Brian Boitano, Grateful Dead drummer Mickey Hart, and other celebrities. They gathered to watch the lights go out, listening to the music of Jason Damato.
- In Kuala Lumpur, Malaysia, the lights of the world's tallest twin towers, the Petronas Towers, were turned off.

Google

Earth Hour 2008 received free publicity from the Google corporation. From 12:00 a.m.

on March 29, 2008 until the end of Earth Hour, the Google homepage in the United States, Colombia, Canada, Denmark, Ireland and the UK was turned to a black background, with the tagline "We've turned the lights out. Now it's your turn – Earth Hour." However, Google stated that for 2009 they would not turn the page black again due to the confusion it caused many users. A common misconception is that having a black background on a web page reduces the power consumption of monitors; LCD monitors use a constant amount of power regardless of which colors are shown. This is not the case for organic LED monitors, though they are not currently in popular use.

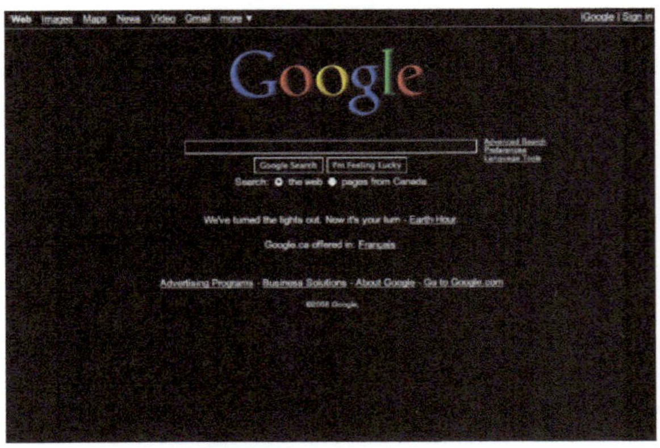

A web screenshot of Google Canada's 'darkened' homepage on March 29, 2008

TV Channels

- Earth Hour was covered extensively in the United States with segments on *The Oprah Winfrey Show*, *NBC Nightly News*, *CBS Evening News*, *The Today Show*, *Good Morning America*, CNN, CNN International, The Weather Channel and more. Several stations around the United States went live with their coverage, including Atlanta NBC station WXIA-TV, who did an hourlong Earth Hour special during the event.
- Canada's The Weather Network moved its studios outside between 8 and 9 pm. EDT for Earth Hour, using only an LED light for the hour.
- *The Agenda with Steve Paikin* on TVOntario ran its full program running only on candlelight.

2009

Earth Hour 2009 was from 8:30 p.m. to 9:30 p.m. local time, March 28, 2009. The campaign was titled "Vote Earth" and was dubbed "the world's first global vote" with one billion votes was the stated aim for Earth Hour 2009, in the context of the pivotal 2009 United Nations Climate Change Conference. WWF reported that 88 countries and 4,159 cities participated in Earth Hour 2009, ten times more cities than Earth Hour 2008 had (2008 saw 400 cities participate).

The global campaign poster for Earth Hour 2009 titled "Vote Earth" created by Shepard Fairey.

Among the participants in 2009 was, for the first time, the United Nations Headquarters in New York City.

In Egypt, the lights went out on the Sphinx and the Great Pyramids of Giza from 8:30 to 9:30 pm.

The Philippines saw participation from 647 cities and towns; over 10 million Filipinos were estimated to have joined in the hour-long lights-off. This was followed by Greece with 484 cities and towns participating, and Australia with 309.

Despite official organisers WWF stating that the event is not about the reduction in electricity, a number of public institutions reported on electricity savings in their cities to see participation numbers. The Canadian province of Ontario, excluding the city of Toronto, saw a decrease of 6% in electricity USge while Toronto saw a decrease of 15.1% (nearly doubled from 8.7% the previous year) as many businesses darkened, including the landmark CN Tower.

The Philippines was able to save 611 MWh of electricity during the time period, which is said to be equivalent to shutting down a dozen coal-fired power plants for an hour.

Swedish electricity operator Svenska Kraftnät recorded a 2.1% decrease in power consumption from its projected figure between 8 p.m. and 9 pm. The following hour, the corresponding number was 5%. This is equivalent to the consumption of approximately half a million households out of the total 4.5 million households in Sweden.

According to Vietnam Electricity Company, Vietnam's electricity demand fell 140 MWh during Earth Hour.

Participation

96 countries and territories on 6 continents participated in the event in 2009.

Participating Television and Radio Stations

- The National Geographic Channel suspended regular programming for an hour and showed how to reduce energy consumption during Earth Hour.
    - National Geographic Channel Asia suspended broadcast on March 28, 2009 from 8:30 p.m. to 9:30 p.m.
- Cartoon Network and Magic 105.4 FM broadcast Earth Hour at 20:30 for the event.

Malaysia's 8TV halted transmission for one hour starting from 20:30

- DhiTV and Villa TV halted transmission for one hour in Maldives from 20:30.
- Canal 5 in Mexico halted transmission for one hour in Mexico City at 20:30.
- Philippine network ABS-CBN turned off the lights in their studio from 20:30 to 21:30.
    - Naga City internet radio stations Zone105 and XFM Naga went offline at 20:30.
- Radiotelevisión Española (RTVE) turned off the lights in their newsrooms and their sets.

2010

The metal structure of the greenhouses of the curitiban Botanic Garden (Curitiba, Paraná, Southern Brazil), with its lights off on March 27, 2010

A picture of 1600 pandas (made in Thailand by the World Wide Fund for Nature) exposed in Boulogne-Billancourt (France) on the occasion of Earth Hour

Earth Hour 2010 was held from 8:30 p.m. to 9:30 p.m. local time on March 27. In Israel, the hour was held on April 22.

126 countries participated in Earth Hour 2010.

In the United States polling showed that an estimated 90,000,000 Americans participated in Earth Hour as lights were turned off around the country, including landmarks such as Mount Rushmore, the Las Vegas Strip, the Empire State Building and Niagara Falls.

Some cities and landmarks took the opportunity to make more long-term adjustments to their everyday power consumption. In Chicago, the Building Owners and Managers Association (BOMA) developed lighting guidelines to reduce light pollution and reduce the carbon footprint of downtown buildings. Mount Rushmore in South Dakota started powering down each night around 9 p.m. instead of 11 p.m.

In Vietnam, electricity demand fell 500,000 kWh during Earth Hour 2010, which was three times larger than the first time the country joined the event in 2009.

In the Philippines, 1,067 towns and cities pledged participation in 2010 and over 15 million Filipinos participated in the event.

About 4000 cities participated, including landmarks such as Big Ben, the Empire State Building, the Sydney Opera House, the Eiffel Tower, the Parthenon, the Brandenburg Gate, and the Forbidden City.

Celebrity Earth Hour ambassadors included South African Archbishop Desmond Tutu, former New Zealand Prime Minister Helen Clark, President of Vietnam Nguyễn Minh Triết, and London Mayor Boris Johnson.

Earth Hour garnered support from many corporations including Coca-Cola Enterprises, Wells Fargo, IKEA, HSBC, PwC, Accenture and Nokia Siemens Networks.

Participating TV Channels and Radio Stations

- National Geographic Channel Asia and Cartoon Network both suspended broadcasting from 8:30 p.m. to 9:30 p.m.
- In the Philippines, GMA Network turned off lights in their building from 8:30 p.m. to 9:30 p.m, while ABS-CBN stopped broadcasting and turned off their lights.
- Vietnam's FBNC channel joined hands with Earth Hour Vietnam.
- *The Agenda with Steve Paikin* on TVOntario ran its full program running only on candlelight again.

Innovative Environmental Media

Australian advertising agency Wunderman Sydney produced a marketing piece to support Earth Hour, their pro-bono client since 2009. To encourage businesses to participate in Earth Hour 2010, five-thousand 'Plant Spikes' were produced and distributed to offices around Australia. The spike was designed to be inserted in office pot plants by plant hire company TPR Group as they serviced plants in businesses nationwide.

To ensure the spike was completely environmentally friendly, the agency partnered with printer STI Lilyfield to invent an organic printer's ink containing natural plant fertilizer derived from *Durvillaea potatorum* and extract of *Ascophyllum nodosum* to promote healthy plant growth and improve resistance to insect and fungal attack. Printed on 100% FSC-certified paper, this ink fertilized the pot plants as the spike biodegraded naturally in the plant's soil.

2011

Earth Hour poster from WWF-China, encouraging people to make lifestyle changes beyond the 60-minute event.

Earth Hour 2011 was the biggest year in the campaign's five-year history, reaffirming it as the largest ever voluntary action for the environment. In 2011, the tagline "Beyond the Hour" was adopted by organisers as a way to encourage people to take their commitment to the cause beyond the 60-minute event. Together with agency Leo Burnett, Earth Hour unveiled an updated planet themed logo that included a small plus symbol to the right of the signature "60" which was used in previous years. The 60+ symbol continues to be the main logo used by campaign organisers around the world.

Earth Hour 2011 took place in a record 5,251 cities and towns in 135 countries and territories on all seven continents. It had an estimated reach of 1.8 billion people across the globe. In addition to this, the campaign's digital footprint grew to 91 million.

In 2011, some of the world's most well-known landmarks, including the Forbidden City, Eiffel Tower, Buckingham Palace, Golden Gate Bridge, Table Mountain, Christ the Redeemer statue and Sydney Opera House switched off their lights for Earth Hour's global "lights out" event.

In India, Earth Hour 2011 was held on March 26, 2011 from 8:30 p.m. to 9:30 pm. IST, flagged off by the Chief Minister of Delhi Sheila Dixit and Earth Hour 2011 Ambassador and Bollywood actress Vidya Balan in the presence of Jim Leape, Director General, WWF International. Rosebowl channel suspended broadcasting from 8.30 p.m. to 9.30 p.m. to mark the observance of Earth Hour.

In Azerbaijan, Maiden Tower darkened for Earth Hour.

The Philippines, which has been an active participant of the Earth Hour, had an early "earth hour" when power was accidentally interrupted, plunging Metro Manila and nearby provinces into darkness. After power was restored, major buildings, commercial centers and residential areas in Metro Manila and most provinces continued to turn off their lights, while participating channels in the Philippines, ABS-CBN and Cartoon Network halted their transmissions for an hour.

30 provinces and cities in Vietnam took part in Earth Hour 2011 with the main event held in Nha Trang. The nation's electricity demand fell 400,000 kWh, one-fifth less than the previous year's. Vietnam managed to save 500 million VND (US$23,809) thanks to the saved power.

YouTube promoted the Earth Hour by changing its logo, and by adding a switch on/off feature near the title of each video, so that users could change the background color from white to black.

One of the least co-operative areas traditionally has been Alberta; in 2008, Calgary's power consumption went up during Earth Hour. The trend continued in 2011 when Edmonton's power usage also increased. While Calgary's power usage went down in

2011 during the event, electricity officials could not distinguish their readings between normal usage and a conscious attempt to participate.

2012

It was announced that the Earth Hour Global headquarters was moving from Sydney to Singapore in February 2012. A launch event took place at ION Orchard on February 20, with the move supported by Singapore's Economic Development Board (EDB) and WWF-Singapore.

Earth Hour 2012 was observed on March 31, 2012 from 8:30 p.m. to 9:30 p.m. (participants' local time). It took place in more than 7000 cities and towns across 152 countries and territories, making it the biggest growth year for the campaign since 2009. It was also the first year that Earth Hour was celebrated in space, with Dutch astronaut André Kuipers tweeting at various moments during the event's trek around the globe.

I Will If You Will

Earth Hour's 2012 campaign poster and tagline Dare the World to Save the Planet".

The global theme for WWF's Earth Hour in 2012 was "I Will If You Will", a two-year campaign aimed to engage the movement's followers to go beyond 60 minutes and to challenge their own networks of friends to become more environmentally conscious throughout the year. Using a YouTube platform, the campaign saw video challenges set by celebrities, politicians, organisations and everyday individuals that were fulfilled when a designated number of accepts were achieved on the platform.

The campaign was supported by supermodel Miranda Kerr, who conducted a free yoga class after 500 fans accepted her challenge, and Greenpeace Executive Director Kumi Naidoo who dyed his beard green at the Rio+20 Conference in South Africa after 10,000 people accepted his challenge to support Earth Hour.

As in the previous year, YouTube changed its logo and added a light switch feature near video titles, so that users could change the background color from white to black. This support helped generate 4.6 million hits on the "I Will If You Will" platform during the week of Earth Hour, with more than 200,000 people personally pledging to take a direct action beyond the 60-minute event.

Environmental Legislation Change in Russia

WWF-Russia generated 122,000 signatures on an Earth Hour petition to strengthen "oil pollution" laws for marine waters of Russia. The petition was entered into the Russian parliament after the March event, and a strengthened law to protect the country's seas was passed in December that year. WWF-Russia had used the "I Will If You Will" challenge idea to generate signatures on the petition, with various Russian celebrities including politician and former boxer Nikolai Valuev challenging the country to sign the petition.

2013

Italy, Verona, Arena with backlight on on square Bra, in the bottom Town Hall before Earth Hour 2013

Italy, Verona, Arena with backlight off on square Bra, in the bottom Town Hall during Earth Hour 2013

Earth Hour 2013 was held across the world on Saturday, March 23 at 8:30 p.m. to 9:30 p.m. local time to avoid taking place after European Summer Time began, ensuring a greater impact for the lights-off event. It was also changed to avoid coinciding with the Christian Holy Saturday, which fell on March 30 of that year.

Globally, there was a continuation of the "I Will If You Will" theme, with certain countries and territories producing video challenges from notable ambassadors to encourage participation in the event and certain environmental actions. Organisers note that in 2013, many countries and territories began producing environmental outcomes through the event.

Africa

In 2013, the world's first Earth Hour Forest began in Uganda, an ongoing project that aims to restore 2700 hectares of degraded land. Standard Chartered Bank-Uganda pledged to help fill the forest with more than 250,000 trees.

Earth Hour commemorations in Madagascar had as their highlight the distribution of one thousand wood-saving stoves to victims of the cyclone Haruna in the southern town of Toliara, extensively damaged in February 22 storm. WWF-Madagascar and ADES (Association pour le Développement de l'Energie Solaire) distributed an additional 2,200 wood-saving stoves later that year.

Former President of Botswana, Festus Mogae promised to plant one million indigenous trees over four years, as part of his "I Will If You Will" challenge for Earth 2013. He planned to kick off his IWIYW challenge by planting 100,000 trees in a severely degraded area in Southern part of the country called Goodhope. The remaining trees were to be planted in other parts of the country which were also in need of land rehabilitation, such as in the North Eastern and Western regions of Botswana.

Europe

WWF-Russia launched its 2013 campaign aiming to secure more than 100,000 signatures from Russian citizens to petition for amendments to the current forest legislation. The petition reached more than 127,000 signatures before the Earth Hour event, ensuring the legislation was debated in the State Duma by politicians.

At the global media launch for Earth Hour 2013, CEO and co-founder Andy Ridley spoke about the movement's significant environmental outcomes beyond the hour: "People from all walks of life, from all nations around the world, are the lifeblood of the Earth Hour interconnected global community. They have proven time and time again that if you believe in something strongly enough, you can achieve amazing things. These stories aren't unique, this is happening all over the world," he said.

Participating Countries

Earth Hour 2013's participants included:

- Spain
- Hong Kong
- Bulgaria

- Samoa
- Fiji
- New Zealand
- Australia, Sydney Opera House and Sydney Harbour Bridge
- Argentina
- Iran, Milad Tower, Saat Tower, Siosepol and Naqsh-e Jahan Square
- New Caledonia
- Papua New Guinea
- Japan, Tokyo Tower
- South Korea
- Philippines
- Indonesia
- Norway
- Malaysia, Petronas Towers
- Taiwan, Taipei 101
- Singapore, Marina Bay Sands
- China, The Bird's Nest (stadium) (Beijing)
- South Africa
- India, Gateway of India
- Dubai, Burj Khalifa
- Bahrain, Bahrain World Trade Center, Al Fateh Grand Mosque, Isa Cultural Centre, Electricity and Water Authority, Bahrain International Circuit and the Al Moayyed Tower
- United Arab Emirates, Shaikh Zayed Grand Mosque and Shaikh Zayed Bridge
- Greece, Acropolis of Athens, White Tower of Thessaloniki
- France, Eiffel Tower
- England, Big Ben, Houses of Parliament

- Italy, David in Florence
- Denmark, Copenhagen's Little Mermaid statue
- Canada, all 30 National Hockey League arenas (7 of which are in Canada), Hamilton Utilities and Horizon Utilities
- United States of America
    - all 30 National Hockey League arenas (23 of which are in the United States)
    - LAX
- Iceland

Participating Organisations

Multinational participation:

- Anschutz Entertainment Group, 58 venues
- Barclays Plc, Barclays Center Brooklyn and several other centers

2014

Earth Hour 2014 took place on Saturday, March 29, during the same 8:30 to 9:30 pm local timeslot. Earth Hour Blue was launched as a global crowdfunding and crowd-sourcing platform for the planet. "It is all about the collective effort of individuals around the world getting together to help fund or add their voice to support on-the-ground environmental and social projects that deliver real outcomes."

The Earth Hour 2014 Report highlighted a broad range of environmental outcomes achieved by the movement across 162 countries and territories around the world. More than US$60,000 was raised on the Earth Hour Blue platform for grassroots environmental projects run by WWF. The movement also saw campaigns to help protect Australia's Great Barrier Reef, the launch of a Blue Sky App in China, and the delivery of thousands of wood efficient stoves to communities in Madagascar.

The event was also supported by the cast and producers of *The Amazing Spider-Man 2*, who attended the Earth Hour event held by WWF-Singapore.

2015

Earth Hour 2015 took place on Saturday, March 28, again between 8:30 and 9:30 pm local time. The tagline for the global campaign was "Change Climate Change", returning to the movement's original focus to initiate citizen action on global warming. A day before the event, over 170 countries and territories had confirmed their participation; with more than 1200 landmarks and close to 40 UNESCO world heritage sites set for the switch off.

For the second year running, Earth Hour Blue aims to raise funds for WWF organised climate focused projects on a crowdfunding platform. This year, crowdfunding projects include solar light distribution in the Philippines and India, and wildlife based projects from Colombia, Uganda and Indonesia.

Uniquely participating in the Earth Hour activity are the inhabitants of an island called Sibuyan in the Philippines who turned on their lights to elevate the message of using renewable energy. The island's source of electricity is a mini-hydro power plant.

2016

Earth Hour 2016 was on Saturday, March 19, from 8:30 p.m. to 9:30 p.m. during participants' local time. It was also changed to avoid coinciding with the Christian Holy Saturday, which fell on March 26 of that year. It was the 10th year anniversary of the campaign's beginnings in Sydney, Australia. Östersund in Sweden cancelled the 2016 event, following a spate of sex attacks, highlighting safety as a subject for discussion when saving resources.Almost all the countries in the world observed Earth Hour, Pakistan also observed this day

Celebrity Endorsements

Earth Hour has been supported around the world by Nelson Mandela, supermodel Miranda Kerr, footballer Lionel Messi, the United Nations Secretary General Ban Ki-moon, Greenpeace International Executive Director Kumi Naidoo, former US Vice President Al Gore, the President of Fiji Ratu Epeli Nailatikau, André Kuipers and the International Space Station, actress Isabel Lucas, TV stars Bill and Giuliana Rancic, Yoko Ono, Stephen Fry, Cate Blanchett, Coldplay, among many others.

In 2014, Earth Hour was supported by Sony Pictures and the cast of *The Amazing Spider-Man 2*, including actors Andrew Garfield, Emma Stone, Jamie Foxx and director Marc Webb. The cast put their support behind Earth Hour Blue crowdfunding projects, asking their fans to raise funds for on-the-ground environmental projects from across the world. *Spider-Man* and its cast and director also attended the WWF-Singapore Earth Hour event in 2014.

Actress and co-founder of The Honest Company, Jessica Alba, was the Earth Hour 2013 global ambassador. Supermodel Miranda Kerr served as global ambassador in 2011 and 2012, helping launch Earth Hour's "I Will If You Will" platform.

Chinese actress Li Bingbing has been a longtime supporter of the campaign, attending WWF-China's Earth Hour events each year. Supermodel Gisele Bündchen is also a longtime supporter of Earth Hour, taking to social media to encourage fans in Brazil and around the world to switch off.

Organisations That Support Earth Hour

Earth Hour is supported around the world by Woodland, CBRE Group, the National Hockey League, FIFA, UEFA, Manchester United, Hilton Worldwide, Girl Scouts of the US, World Organisation of the Scouts Movement, UNESCO, the UN Environment Programme, the International Trade Union Confederation, HSBC, World Association of the Girl Guides and Girl Scouts, Philips, Ikea, The Body Shop, ING Vysya Bank and more.

Measurement of Reduction in Electricity Use

The Earth Hour Global FAQ page states:

Earth Hour does not purport to be an energy/carbon reduction exercise, it is a symbolic action. Therefore, we do not engage in the measurement of energy/carbon reduction levels for the hour itself. Earth Hour is an initiative to encourage individuals, businesses and governments around the world to take accountability for their ecological footprint and engage in dialogue and resource exchange that provides real solutions to our environmental challenges. Participation in Earth Hour symbolises a commitment to change beyond the hour.

A 2014 study published in *Energy Research and Social Science* compiled 274 measurements of observed changes in electricity demand caused by Earth Hour in 10 countries, spanning 6 years, and found that the events reduced electricity consumption an average of 4%. The study noted the policy challenge of converting Earth Hour's short-term energy saving into longer-term actions, including sustained changes in behavior and investment.

Criticism

Some critics point out that the reduction in power consumption in most cases is insignificant. *The Herald Sun* equated the power savings in the Sydney central business district to "taking 48,613 cars off the road for 1 hour".Australian columnist Andrew Bolt pointed out that "A cut so tiny is trivial – equal to taking six cars off the road for a year".

Other criticisms of Earth Hour have included the following:

- The Competitive Enterprise Institute has introduced an opposing *Human Achievement Hour* in celebration of human progress in various fields of industry, including technology, medicine, energy, and more. During this hour, the Institute suggests that people celebrate by using modern technology such as electricity, telecommunications and indoor plumbing.

- Bjørn Lomborg, author of *The Skeptical Environmentalist*, wrote, "It is vital to make solar and other new technology cheaper than fossil fuels quickly so we can turn off carbon energy sources for a lot longer than one hour and keep the planet running... Fossil fuels literally gave us an enlightenment, by lighting our world and giving us protection from the fury of the elements. It is ironic that today's pure symbolism should hark back to a darker age."

- Bjørn also pointed out the feel-good factor Earth Hour creates, noting that it is an "ineffective feel good event" that makes people feel they are doing something for the environment, while in reality the amount of carbon emissions reduced by the earth hour is negligible.

- *The Christian Science Monitor* said that most candles are made from paraffin, a heavy hydrocarbon derived from crude oil, a fossil fuel, and that depending on how many candles a person burns (if one uses candles during Earth Hour), whether or not they normally use compact fluorescent light bulbs, and what source of energy is used to produce their electricity, in some cases, replacing light bulbs with candles will cause an increase, instead of a decrease, in carbon dioxide emissions.

- On March 29, 2009, one day after Earth Hour 2009, *Dân Trí Daily News* published an editorial expressing concern that many young people chose to drive around the darkened city of Hanoi for fun, exhausting petroleum instead of electricity and resulting in long traffic jams.

- In 2009, economist Ross McKitrick criticized the idea, saying, "Abundant, cheap electricity has been the greatest source of human liberation in the 20th century.[...] The whole mentality around Earth Hour demonizes electricity."

- In March 2010, *The Daily Telegraph* quoted Ross Hayman, head of media relations at the UK National Grid, as saying "it could therefore result in an increase in carbon emissions" due to complications related to rapidly lowering then raising electricity generation.

- In February 2010, Rick Giles, president of ACT on Campus, the youth wing of New Zealand's ACT Party, appeared on the morning television show *Sunrise* to denounce Earth Hour and instead suggested the celebration of "Edison Hour". He argued that Earth Hour is an "anti-technology" cause, and that people will simply use candles instead, which is undesirable as they are petroleum-based. He argued that if we are heading for some kind of disaster, it makes sense to use technology to combat this. Rick said "I think my argument is so powerful that it's not necessary to talk about it".

- The Ayn Rand Institute wrote, "Participants spend an enjoyable sixty minutes in the dark, safe in the knowledge that the life-saving benefits of industrial civilization are just a light switch away... Forget one measly hour with just the lights off. How about Earth Month... Try spending a month shivering in the dark without heating, electricity, refrigeration; without power plants or generators; without any of the labor-saving, time-saving, and therefore life-saving products that industrial energy makes possible."

- Expressing sarcastic support for Earth Hour, the pro-carbon Carbon Sense Coalition wants Earth Hour to be renamed "Blackout Night", and to be held outside on the shortest and coldest day of the year "...to prepare our population for the dark days ahead".
- During the 2010 Earth Hour in the city of Uusikaupunki in Finland, a 17-year-old female motorcyclist hit a 71-year-old man, who was walking on the street instead of the sidewalk for an unknown reason. The man died from his injuries, while the motorcyclist and her passenger were uninjured. At the time of the accident the street lights had been turned off as part of the Earth Hour. The police stated that the lack of street lighting may have played a part in the accident, while the mayor believed the city's street lights would have been too dim to prevent it even if they had been on.
- Jeremy Clarkson, ex-host of the BBC motoring programme *Top Gear*, claimed switching on all electrical items in his home as a protest against the perceived impact of Earth Hour, claiming the event would have little to no effect on attitudes towards climate change.

Energy Hierarchy

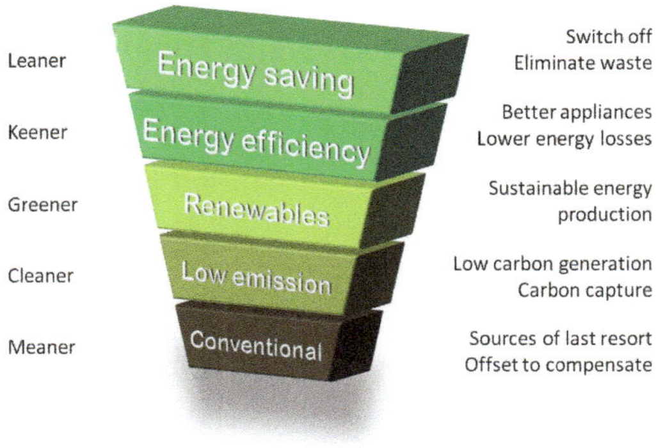

The Energy Hierarchy with the most favoured options at the top

The Energy Hierarchy is a classification of energy options, prioritised to assist progress towards a more sustainable energy system. It is a similar approach to the waste hierarchy for minimising resource depletion, and adopts a parallel sequence.

The highest priorities cover the prevention of unnecessary energy usage both through eliminating waste and improving energy efficiency. The sustainable production of energy resources is the next priority. Depletive and waste-producing energy generation options are the lowest priority.

For an energy system to be sustainable: the resources applied to producing the energy must be capable of lasting indefinitely; energy conversion should produce no harmful by-products, including net emissions, nor wastes which cannot be fully recycled; and it must be capable of meeting reasonable energy demands.

Energy Saving

The top priority under the Energy Hierarchy is energy conservation or the prevention of unnecessary use of energy. This category includes eliminating waste by turning off unneeded lights and appliances and by avoiding unnecessary journeys. Heat loss from buildings is a major source of energy wastage, so improvements to building insulation and air-tightness can make a significant contribution to energy conservation.

Many countries have agencies to encourage energy saving.

Energy Efficiency

The second priority under the energy hierarchy is to ensure that energy that is used is produced and consumed efficiently. Energy efficiency has two main aspects.

Conversion Efficiency of Energy Consumption

Energy efficiency is the ratio of the productive output of a device to the energy it consumes.

Energy efficiency was a lower priority when energy was cheap and awareness of its environmental impact was low. In 1975 the average fuel economy of a car in the US was under 15 miles per gallon Incandescent light bulbs, which were the most common type until the late 20th century, waste 90% of their energy as heat, with only 10% converted to useful light.

More recently, energy efficiency has become a priority. The last reported average fuel efficiency of US cars had almost doubled from the 1975 level; LED lighting is now being promoted which are between five and ten times more efficient than incandescents. Many household appliances are now required to display labels to show their energy efficiency.

Conversion Efficiency of Energy Production

Losses are incurred when energy is harvested from the natural resource from which it is derived, such as fossil fuels, radioactive materials, solar radiation or other sources.

Most electricity production is in thermal power stations, where much of the source energy is lost as heat. The average efficiency of world electricity production in 2009 was c.37%.

A priority in the Energy Hierarchy is to improve the efficiency of energy conversion, whether in traditional power stations or by improving the performance ratio of Photovoltaic power stations and other energy sources.

Overall efficiency and sustainability can also be improved by capacity- or fuel-switching from less efficient, less sustainable resources to better ones; but this is mainly covered under the fourth level of the hierarchy.

Sustainable Energy Production

Renewable energy describes naturally occurring, theoretically inexhaustible sources of energy. These sources are treated as being inexhaustible, or naturally replenished, and fall into two classes.

Elemental Renewables

The first class of renewables derive from climatic or elemental sources, such as sunlight, wind, waves, tides or rainfall (hydro power). Geothermal energy from the heat of the earth's core also falls in this category.

These are treated as being inexhaustible because most derive ultimately from energy emanating from the sun, which has an estimated life of 6.5 billion years.

Bio-energy

The other main class of renewables, bioenergy, derives from biomass, where the relatively short growing cycle means that usage is replenished by new growth. Bioenergy is usually converted by combustion, and therefore gives rise to carbon emissions. It is treated as carbon neutral overall, because an equivalent amount of carbon dioxide will have been extracted from the atmosphere during the growing cycle.

Bioenergy sources can be solid, such as wood and energy crops; liquid, such as biofuels; or gaseous, such as biomethane from anaerobic digestion.

Low Impact Energy Production

The next priority in the hierarchy covers energy sources that are not entirely sustainable, but have a low environmental impact. These include the use of fossil fuels with carbon capture and storage.

Nuclear energy is sometimes treated as a low impact source, because it has low carbon emissions.

High Impact Energy Production

The lowest priority under the energy hierarchy is energy production using unsustainables sources, such as unabated fossil fuels. Some also place nuclear energy in this category, rather than the one above, because of the required management/storage of highly hazardous radioactive waste over extremely long (hundreds of thousands of years or more) timeframes and depletion of uranium resources.

There is a consensus that the share of such energy sources must decline.

Within this tier, there are possibilities for limiting adverse impacts by switching from the most damaging fuel sources, such as coal, to less emittive sources, such as gas.

Many suggest that when such high impact energy usage has been minimised, the effects of any unavoidable residual usage should be counterbalanced by emissions offsetting.

Origins of the Energy Hierarchy

The Energy Hierarchy was first proposed in 2005 by Philip Wolfe, when he was Director General of the Renewable Energy Association. This first version had three levels; energy efficiency, renewables and traditional energy production. It was endorsed and adopted in 2006 by a consortium of institutions, associations and other bodies in the Sustainable Energy Manifesto. Subsequently the concept has been adopted and refined by others in the energy industry and in government.

One Watt Initiative

The One Watt Initiative (or the proper One-Watt Initiative) is an energy-saving initiative by the International Energy Agency (IEA) to reduce standby power-use by any appliance to not more than one watt in 2010, and 0.5 watts in 2013, which has given rise to regulations in many countries and regions.

Standby Power

Standby power, informally called *vampire* or *phantom power*, refers to the electricity consumed by many appliances when they are switched off or in standby mode. The typical standby power per appliance is low (typically from less than 1 to 25 W), but, when multiplied by the billions of appliances in houses and in commercial buildings, standby losses represent a significant fraction of total world electricity use. According to Alan Meier, a staff scientist at the Lawrence Berkeley National Laboratory, standby power before the One Watt Initiative proposals were implemented as regulations accounted for as much as 10% of household power consumption. A study in France found that standby power accounted for 7% of total residential consumption, and other studies put the proportion of consumption due to standby power at 13%.

The IEA estimated in 2007 that standby produced 1% of the world's carbon dioxide (CO_2) emissions. To put the figure into context, total air travel contributes less than 3% of global CO_2 emissions.

Standby power can be reduced by technological means, reducing power used without affecting functionality, and by changing users' operating procedures.

Policy

The One Watt Initiative was launched by the IEA in 1999 to ensure through international cooperation that by 2010 all new appliances sold in the world use only one watt in standby mode. This would reduce CO_2 emissions by 50 million tons in the OECD countries alone by 2010; the equivalent to removing 18 million cars from the roads.

In 2001 US President George W. Bush issued Executive Order 13221, which states that every government agency, "when it purchases commercially available, off-the-shelf products that use external standby power devices, or that contain an internal standby power function, shall purchase products that use no more than one watt in their standby power consuming mode."

By 2005 South Korea and Australia had introduced the one watt benchmark in all new electrical devices, and according to the IEA other countries, notably Japan and China, had undertaken "strong measures" to reduce standby power use.

In July 2007 California's 2005 appliance standards came into effect, limiting external power supply standby power to 0.5 watts.

On 6 January 2010, the European Commission's EC Regulation 1275/2008 came into force regulating requirements for standby and "off mode" electric power consumption of electrical and electronic household and office equipment. The regulations mandate that from 6 January 2010 "off mode" and standby power shall not exceed 1W, "standby-plus" power (providing information or status display in addition to possible reactivation function) shall not exceed 2W (these figures are halved on 6 January 2013). Equipment must, where appropriate, provide off mode and/or standby mode when the equipment is connected to the mains power source.

Hubbert Peak Theory

The Hubbert peak theory says that for any given geographical area, from an individual oil-producing region to the planet as a whole, the rate of petroleum production tends to follow a bell-shaped curve. It is one of the primary theories on peak oil.

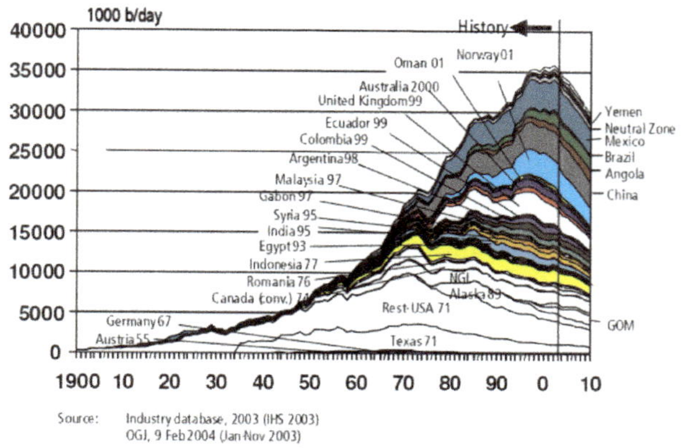

2004 U.S. government predictions for oil production other than in OPEC and the former Soviet Union

Choosing a particular curve determines a point of maximum production based on discovery rates, production rates and cumulative production. Early in the curve (pre-peak), the production rate increases because of the discovery rate and the addition of infrastructure. Late in the curve (post-peak), production declines because of resource depletion.

The Hubbert peak theory is based on the observation that the amount of oil under the ground in any region is finite, therefore the rate of discovery which initially increases quickly must reach a maximum and decline. In the US, oil extraction followed the discovery curve after a time lag of 32 to 35 years. The theory is named after American geophysicist M. King Hubbert, who created a method of modeling the production curve given an assumed ultimate recovery volume.

Hubbert's Peak

"Hubbert's peak" can refer to the peaking of production of a particular area, which has now been observed for many fields and regions.

Hubbert's Peak was thought to have been achieved in the United States contiguous 48 states (that is, excluding Alaska) in the early 1970s. Oil production peaked at 10,200,000 barrels per day (1,620,000 m^3/d) and then declined for several years since. Yet, recent advances in extraction technology, particularly those that led to the extraction of tight oil and oil from shale,have drastically changed the picture. A decline in production followed the 1970s peak, but it has been succeeded by a major increase in production.

Peak oil as a proper noun, or "Hubbert's peak" applied more generally, refers to a predicted event: the peak of the entire planet's oil production. After Peak Oil, according to the Hubbert Peak Theory, the rate of oil production on Earth would enter a terminal decline. On the basis of his theory, in a paper he presented to the American Petroleum Institute in 1956, Hubbert correctly predicted that production of oil from conventional sources would peak in the continental United States around 1965–1970. His prediction of inevitable decline has been incorrect, but the 1970 peak has yet not been surpassed. Hubbert further predicted a worldwide peak at "about half a century" from publication and approximately 12 gigabarrels (GB) a year in magnitude. In a 1976 TV interview Hubbert added that the actions of OPEC might flatten the global production curve but this would only delay the peak for perhaps 10 years. The development of new technologies has provided access to large quantities of unconventional resources, and the boost of production has largely discounted Huppert's prediction. In the future, pressure to limit the use of fossil fuels (and so reduce the release of greenhouse gasses) will curb production, not exhaustion of resources.

Hubbert's Theory

Hubbert Curve

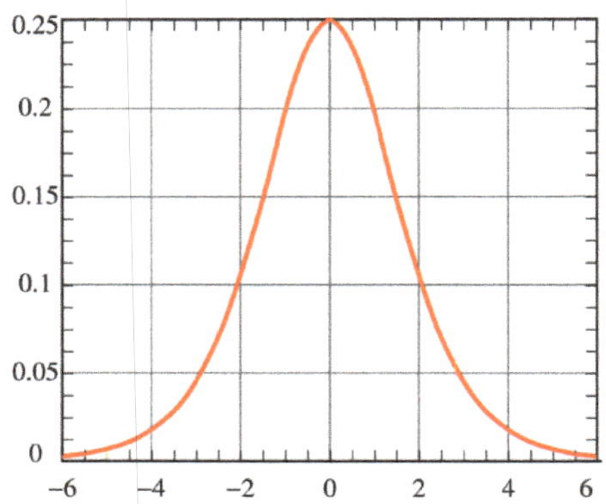

The standard Hubbert curve. For applications, the x and y scales are replaced by time and production scales.

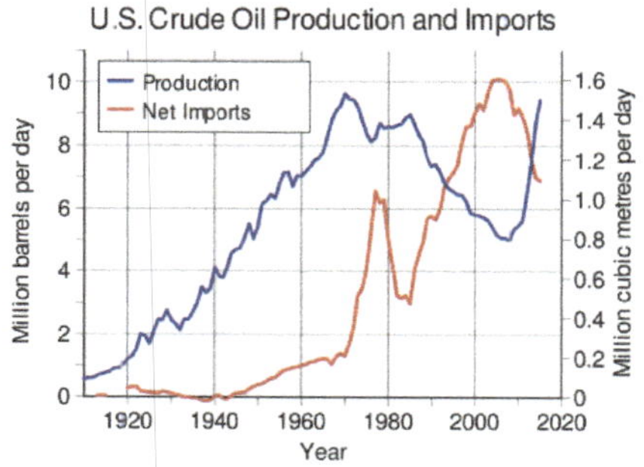

U.S. Oil Production and Imports 1910 to 2012

In 1956, Hubbert proposed that fossil fuel production in a given region over time would follow a roughly bell-shaped curve without giving a precise formula; he later used the Hubbert curve, the derivative of the logistic curve, for estimating future production using past observed discoveries.

Hubbert assumed that after fossil fuel reserves (oil reserves, coal reserves, and natural gas reserves) are discovered, production at first increases approximately exponentially, as more extraction commences and more efficient facilities are installed. At some point, a peak output is reached, and production begins declining until it approximates an exponential decline.

The Hubbert curve satisfies these constraints. Furthermore, it is roughly symmetrical, with the peak of production reached when about half of the fossil fuel that will ultimately be produced has been produced. It also has a single peak.

Given past oil discovery and production data, a Hubbert curve that attempts to approximate past discovery data may be constructed and used to provide estimates for future production. In particular, the date of peak oil production or the total amount of oil ultimately produced can be estimated that way. Cavallo defines the Hubbert curve used to predict the U.S. peak as the derivative of:

$$Q(t) = \frac{Q_{\max}}{1 + ae^{-bt}}$$

where $Q_{\max}$ is the total resource available (ultimate recovery of crude oil), $Q(t)$ the cumulative production, and a and b are constants. The year of maximum annual production (peak) is:

$$t_{\max} = \frac{1}{b}\ln(a).$$

so now the cumulative production $Q(t)$ reaches the half of the total available resource:

$$Q(t) = Q_{\max} >$$

Use of Multiple Curves

The sum of multiple Hubbert curves, a technique not developed by Hubbert himself, may be used in order to model more complicated real life scenarios.

Reliability

Crude Oil

Hubbert, in his 1956 paper, presented two scenarios for US crude oil production:

- most likely estimate: a logistic curve with a logistic growth rate equal to 6%, an ultimate resource equal to 150 Giga-barrels (Gb) and a peak in 1965. The size of the ultimate resource was taken from a synthesis of estimates by well-known oil geologists and the US Geological Survey, which Hubbert judged to be the most likely case.

- upper-bound estimate: a logistic curve with a logistic growth rate equal to 6% and ultimate resource equal to 200 Giga-barrels and a peak in 1970.

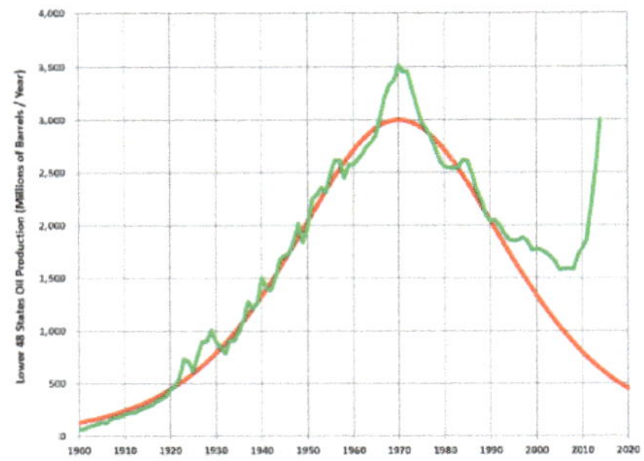

Hubbert's upper-bound prediction for US crude oil production (1956), and actual lower-48 states production through 2014

Hubbert's upper-bound estimate, which he regarded as optimistic, accurately predicted that US oil production would peak in 1970, although the actual peak was 17% higher than Hubbert's curve. Production declined, as Hubbert had predicted, and stayed within 10 percent of Hubbert's predicted value from 1974 through 1994; since then, actual production has been significantly greater than the Hubbert curve. The development of new technologies has provided access to large quantities of unconventional resources, and the boost of production has largely discounted Huppert's prediction. In the future, pressure to limit the use of fossil fuels (and so reduce the release of greenhouse gasses) will curb production, not exhaustion of resources.

Hubbert's 1956 production curves depended on geological estimates of ultimate recoverable oil resources, but he was dissatisfied by the uncertainty this introduced, given the various estimates ranging from 110 billion to 590 billion barrels for the US. Starting in his 1962 publication, he made his calculations, including that of ultimate recovery, based only on mathematical analysis of production rates, proved reserves, and new discoveries, independent of any geological estimates of future discoveries. He concluded that the ultimate recoverable oil resource of the contiguous 48 states was 170 billion barrels, with a production peak in 1966 or 1967. He considered that because his model incorporated past technical advances, that any future advances would occur at the same rate, and were also incorporated. Hubbert continued to defend his calculation of 170 billion barrels in his publications of 1965 and 1967, although by 1967 he had moved the peak forward slightly, to 1968 or 1969.

A post-hoc analysis of peaked oil wells, fields, regions and nations found that Hubbert's model was the "most widely useful" (providing the best fit to the data), though many areas studied had a sharper "peak" than predicted.

A 2007 study of oil depletion by the UK Energy Research Centre pointed out that there is no theoretical and no robust practical reason to assume that oil production will follow a logistic curve. Neither is there any reason to assume that the peak will occur when half the ultimate recoverable resource has been produced; and in fact, empirical evidence appears to contradict this idea. An analysis of a 55 post-peak countries found that the average peak was at 25 percent of the ultimate recovery.

Natural Gas

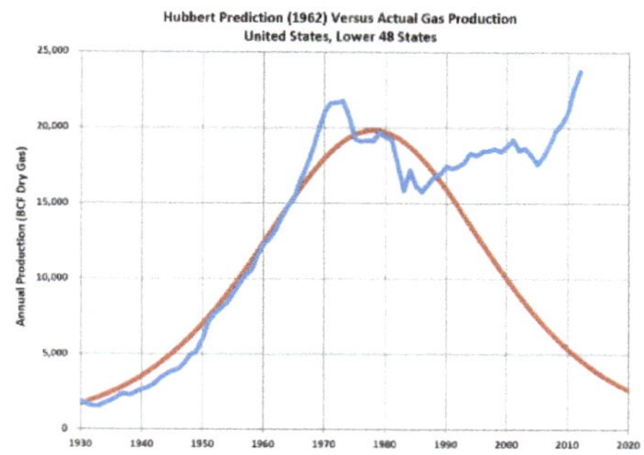

Hubbert's 1962 prediction of US lower 48-state gas production, versus actual production through 2012

Hubbert also predicted that natural gas production would follow a logistic curve similar to that of oil. At right is his gas production curve for the United States, published in 1962.

Economics

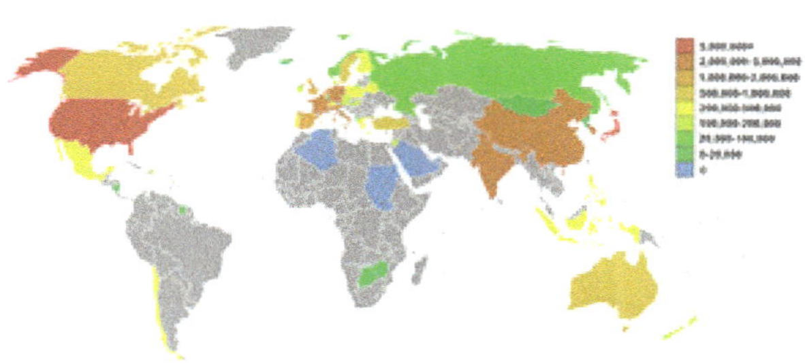

Oil imports by country Pre-2006

Energy Return on Energy Investment

The ratio of energy extracted to the energy expended in the process is often referred to as the Energy Return on Energy Investment (EROI or EROEI). As the EROEI drops to one, or equivalently the Net energy gain falls to zero, the oil production is no longer a net energy source. This happens long before the resource is physically exhausted.

Note that it is important to understand the distinction between a barrel of oil, which is a measure of oil, and a barrel of oil equivalent (BOE), which is a measure of energy. Many sources of energy, such as fission, solar, wind, and coal, are not subject to the same near-term supply restrictions that oil is. Accordingly, even an oil source with an EROEI of 0.5 can be usefully exploited if the energy required to produce that oil comes from a cheap and plentiful energy source. Availability of cheap, but hard to transport, natural gas in some oil fields has led to using natural gas to fuel enhanced oil recovery. Similarly, natural gas in huge amounts is used to power most Athabasca tar sands plants. Cheap natural gas has also led to ethanol fuel produced with a net EROEI of less than 1, although figures in this area are controversial because methods to measure EROEI are in debate.

Advances in technology or experience can lead to greater productivity. The US Energy Information Administration has reported that drilling for shale gas and light tight oil in the United States became much more efficient throughout the period 2007–2014. In terms of oil produced per day of rig drilling time, Bakken wells drilled in January 2014 produced 2.4 times as much oil as those drilled five years earlier, in January 2009. In the Marcellus Gas Trend, wells drilled in January 2014 produced more than nine times as much gas per day of drilling rig time as those drilled five years previously, in January 2009.

Growth-based Economic Models

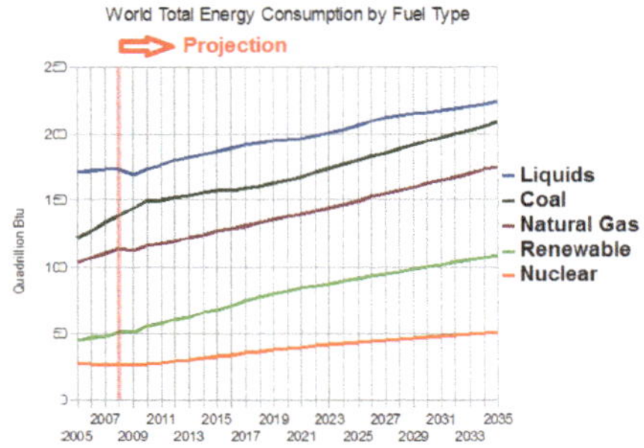

World energy consumption & predictions, 2005–2035. *Source: International Energy Outlook 2011.*

Insofar as economic growth is driven by oil consumption growth, post-peak societies must adapt. Hubbert believed:

Some economists describe the problem as uneconomic growth or a false economy. At the political right, Fred Ikle has warned about "conservatives addicted to the Utopia of Perpetual Growth". Brief oil interruptions in 1973 and 1979 markedly slowed—but did not stop—the growth of world GDP.

Between 1950 and 1984, as the Green Revolution transformed agriculture around the globe, world grain production increased by 250%. The energy for the Green Revolution was provided by fossil fuels in the form of fertilizers (natural gas), pesticides (oil), and hydrocarbon fueled irrigation.

David Pimentel, professor of ecology and agriculture at Cornell University, and Mario Giampietro, senior researcher at the National Research Institute on Food and Nutrition (INRAN), place in their study *Food, Land, Population and the U.S. Economy* the maximum U.S. population for a sustainable economy at 200 million. To achieve a sustainable economy world population will have to be reduced by two-thirds, says the study. Without population reduction, this study predicts an agricultural crisis beginning in 2020, becoming critical c. 2050. The peaking of global oil along with the decline in regional natural gas production may precipitate this agricultural crisis sooner than generally expected. Dale Allen Pfeiffer claims that coming decades could see spiraling food prices without relief and massive starvation on a global level such as never experienced before.

Hubbert Peaks

Although Hubbert peak theory receives most attention in relation to peak oil production, it has also been applied to other natural resources.

Natural Gas

Doug Reynolds predicted in 2005 that the North American peak would occur in 2007. Bentley predicted a world "decline in conventional gas production from about 2020".

Coal

Although observers believe that peak coal is significantly further out than peak oil, Hubbert studied the specific example of anthracite in the USA, a high grade coal, whose production peaked in the 1920s. Hubbert found that anthracite matches a curve closely. Hubbert had recoverable coal reserves worldwide at 2.500×10^{12} metric tons and peaking around 2150 (depending on usage).

More recent estimates suggest an earlier peak. *Coal: Resources and Future Production* (PDF 630KB), published on April 5, 2007 by the Energy Watch Group (EWG), which reports to the German Parliament, found that global coal production could peak in as few as 15 years. Reporting on this, Richard Heinberg also notes that the date of peak annual energetic extraction from coal is likely to come earlier than the date of peak in quantity of coal (tons per year) extracted as the most energy-dense types of coal have been mined most extensively. A second study, *The Future of Coal* by B. Kavalov and S. D. Peteves of the Institute for Energy (IFE), prepared for European Commission Joint Research Centre, reaches similar conclusions and states that ""coal might not be so abundant, widely available and reliable as an energy source in the future".

Work by David Rutledge of Caltech predicts that the total of world coal production will amount to only about 450 gigatonnes. This implies that coal is running out faster than usually assumed.

Fissionable Materials

In a paper in 1956, after a review of US fissionable reserves, Hubbert notes of nuclear power:

> “There is promise, however, provided mankind can solve its international problems and not destroy itself with nuclear weapons, and provided world population (which is now expanding at such a rate as to double in less than a century) can somehow be brought under control, that we may at last have found an energy supply adequate for our needs for at least the next few centuries of the “foreseeable future.””

As of 2012, the identified resources of uranium are sufficient to provide more than 120 years of supply at the present rate of consumption. Technologies such as the thorium fuel cycle, reprocessing and fast breeders can, in theory, extend the life of uranium reserves from hundreds to thousands of years.

Caltech physics professor David Goodstein stated in 2004 that

> “... you would have to build 10,000 of the largest power plants that are feasible by engineering standards in order to replace the 10 terawatts of fossil fuel we're burning today ... that's a staggering amount and if you did that, the known reserves of uranium would last for 10 to 20 years at that burn rate. So, it's at best a bridging technology ... You can use the rest of the uranium to breed plutonium 239 then we'd have at least 100 times as much fuel to use. But that means you're making plutonium, which is an extremely dangerous thing to do in the dangerous world that we live in.”

Helium

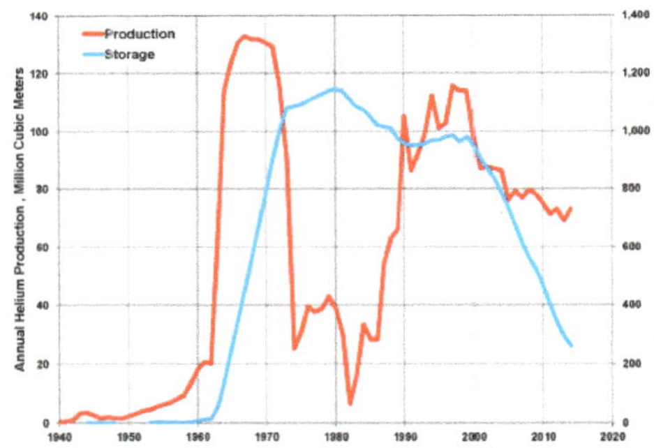

Helium production and storage in the United States, 1940–2014 (data from USGS)

Almost all helium on Earth is a result of radioactive decay of uranium and thorium. Helium is extracted by fractional distillation from natural gas, which contains up to 7% helium. The world's largest helium-rich natural gas fields are found in the United States, especially in the Hugoton and nearby gas fields in Kansas, Oklahoma, and Texas. The extracted helium is stored underground in the National Helium Reserve near Amarillo, Texas, the self-proclaimed “Helium Capital of the World”. Helium production is expected to decline along with natural gas production in these areas.

Helium, which is the second-lightest chemical element, will rise to the upper layers of Earth's atmosphere, where it can forever break free from Earth's gravitational attraction. Approximately 1,600 tons of helium are lost per year as a result of atmospheric escape mechanisms.

Transition Metals

Hubbert applied his theory to "rock containing an abnormally high concentration of a given metal" and reasoned that the peak production for metals such as copper, tin, lead, zinc and others would occur in the time frame of decades and iron in the time frame of two centuries like coal. The price of copper rose 500% between 2003 and 2007 and was attributed by some to peak copper. Copper prices later fell, along with many other commodities and stock prices, as demand shrank from fear of a global recession. Lithium availability is a concern for a fleet of Li-ion battery using cars but a paper published in 1996 estimated that world reserves are adequate for at least 50 years. A similar prediction for platinum use in fuel cells notes that the metal could be easily recycled.

Precious Metals

In 2009, Aaron Regent president of the Canadian gold giant Barrick Gold said that global output has been falling by roughly one million ounces a year since the start of the decade. The total global mine supply has dropped by 10pc as ore quality erodes, implying that the roaring bull market of the last eight years may have further to run. "There is a strong case to be made that we are already at 'peak gold'," he told The Daily Telegraph at the RBC's annual gold conference in London. "Production peaked around 2000 and it has been in decline ever since, and we forecast that decline to continue. It is increasingly difficult to find ore," he said.

Ore grades have fallen from around 12 grams per tonne in 1950 to nearer 3 grams in the US, Canada, and Australia. South Africa's output has halved since peaking in 1970. Output fell a further 14 percent in South Africa in 2008 as companies were forced to dig ever deeper – at greater cost – to replace depleted reserves.

World mined gold production has peaked four times since 1900: in 1912, 1940, 1971, and 2001, each peak being higher than previous peaks. The latest peak was in 2001, when production reached 2,600 metric tons, then declined for several years. Production started to increase again in 2009, spurred by high gold prices, and achieved record new highs each year in 2012, 2013, and in 2014, when production reached 2,990 tonnes.

Phosphorus

Phosphorus supplies are essential to farming and depletion of reserves is estimated at somewhere from 60 to 130 years. According to a 2008 study, the total reserves of phosphorus are estimated to be approximately 3,200 MT, with a peak production at 28 MT/year in 2034. Individual countries' supplies vary widely; without a recycling initiative America's supply is estimated around 30 years. Phosphorus supplies affect agricultural output which in turn limits alternative fuels such as biodiesel and ethanol. Its increasing price and scarcity (global price of rock phosphate rose 8-fold in the 2 years to mid

2008) could change global agricultural patterns. Lands, perceived as marginal because of remoteness, but with very high phosphorus content, such as the Gran Chaco may get more agricultural development, while other farming areas, where nutrients are a constraint, may drop below the line of profitability.

Peak Water

Hubbert's original analysis did not apply to renewable resources. However, over-exploitation often results in a Hubbert peak nonetheless. A modified Hubbert curve applies to any resource that can be harvested faster than it can be replaced.

For example, a reserve such as the Ogallala Aquifer can be mined at a rate that far exceeds replenishment. This turns much of the world's underground water and lakes into finite resources with peak usage debates similar to oil. These debates usually center around agriculture and suburban water usage but generation of electricity from nuclear energy or coal and tar sands mining mentioned above is also water resource intensive. The term fossil water is sometimes used to describe aquifers whose water is not being recharged.

Renewable Resources

- Fisheries: At least one researcher has attempted to perform Hubbert linearization (Hubbert curve) on the whaling industry, as well as charting the transparently dependent price of caviar on sturgeon depletion. The Atlantic northwest cod fishery was a renewable resource, but the numbers of fish taken exceeded the fish's rate of recovery. The end of the cod fishery does match the exponential drop of the Hubbert bell curve. Another example is the cod of the North Sea.
- Air/oxygen: Half the world's oxygen is produced by phytoplankton. The numbers of plankton have dropped by 40% since the 1950s.

Criticisms of Peak Oil

Economist Michael Lynch argues that the theory behind the Hubbert curve is too simplistic and relies on an overly Malthusian point of view. Lynch claims that Campbell's predictions for world oil production are strongly biased towards underestimates, and that Campbell has repeatedly pushed back the date.

Leonardo Maugeri, vice president of the Italian energy company Eni, argues that nearly all of peak estimates do not take into account unconventional oil even though the availability of these resources is significant and the costs of extraction and processing, while still very high, are falling because of improved technology. He also notes that the recovery rate from existing world oil fields has increased from about 22% in 1980 to 35% today because of new technology and predicts this trend will continue. The ratio between proven oil reserves and current production has constantly improved, passing from 20 years in 1948 to 35 years in 1972 and reaching about 40 years in 2003. These improvements occurred even with low investment in new exploration and upgrading technology because of the low oil prices during the last 20 years. However, Maugeri feels that encouraging more exploration will require relatively high oil prices.

Edward Luttwak, an economist and historian, claims that unrest in countries such as Russia, Iran and Iraq has led to a massive underestimate of oil reserves. The Association for the Study of Peak Oil and Gas (ASPO) responds by claiming neither Russia nor Iran are troubled by unrest currently, but Iraq is.

Cambridge Energy Research Associates authored a report that is critical of Hubbert-influenced predictions:

> “Despite his valuable contribution, M. King Hubbert's methodology falls down because it does not consider likely resource growth, application of new technology, basic commercial factors, or the impact of geopolitics on production. His approach does not work in all cases-including on the United States itself-and cannot reliably model a global production outlook. Put more simply, the case for the imminent peak is flawed. As it is, production in 2005 in the Lower 48 in the United States was 66 percent higher than Hubbert projected.”

CERA does not believe there will be an endless abundance of oil, but instead believes that global production will eventually follow an “undulating plateau” for one or more decades before declining slowly, and that production will reach 40 Mb/d by 2015.

Alfred J. Cavallo, while predicting a conventional oil supply shortage by no later than 2015, does not think Hubbert's peak is the correct theory to apply to world production.

Criticisms of Peak Element Scenarios

Although M. King Hubbert himself made major distinctions between decline in petroleum production versus depletion (or relative lack of it) for elements such as fissionable uranium and thorium, some others have predicted peaks like peak uranium and peak phosphorus soon on the basis of published reserve figures compared to present and future production. According to some economists, though, the amount of proved reserves inventoried at a time may be considered “a poor indicator of the total future supply of a mineral resource.”

As some illustrations, tin, copper, iron, lead, and zinc all had both production from 1950 to 2000 and reserves in 2000 much exceed world reserves in 1950, which would be impossible except for how “proved reserves are like an inventory of cars to an auto dealer” at a time, having little relationship to the actual total affordable to extract in the future. In the example of peak phosphorus, additional concentrations exist intermediate between 71,000 Mt of identified reserves (USGS) and the approximately 30,000,000,000 Mt of other phosphorus in Earth's crust, with the average rock being 0.1% phosphorus, so showing decline in human phosphorus production will occur soon would require far more than comparing the former figure to the 190 Mt/year of phosphorus extracted in mines (2011 figure).

References

- Sovacool, Benjamin (2011). Contesting the Future of Nuclear Power. Singapore: World Scientific. p. 308. ISBN 978-981-4322-75-1.
- "Home - WWF Earth Hour: 19 March 2016 - South Africa". WWF Earth Hour: 19 March 2016 - South Africa. Retrieved March 20, 2016.
- "Earth Hour Targets 2014 to be the Crowdfunding Center For Innovation and Creativity for the Planet". crowdfundinsider.com. Retrieved May 31, 2015.
- "Earth Hour Blue: mobilising crowdfunding to support conservation and sustainable development". Green Africa Directory. Retrieved May 31, 2015.
- "SOSSebangau: Seruan Aksi Peduli Kebakaran Hutan Kalimantan". WWF Indonesia (in Indonesian). Retrieved May 31, 2015.
- Timon Singh. "WWF Recruits Spider-Man as Earth Hour's First Superhero Ambassador". inhabitat.com. Retrieved May 31, 2015.
- "Woodland brings 'Earth Hour' initiative to Bengaluru in association with WWF". This Week Bangalore. March 25, 2013. Retrieved 2015-10-05.
- "CBRE Announces Client Properties Totaling 665 Million Square Feet Participated in WWF Earth Hour 2013". CBRE Group. March 26, 2013. Retrieved 2015-10-05.
- James McLaughlin. "Landmarks switch-off the lights in support of Earth Hour's climate change awareness". Mancunian Matters. 2012-03-30. Retrieved on 2015-10-05.
- "Hilton Worldwide Celebrates its Commitment to Living Sustainably with Earth Hour Projects around the World". hiltonworldwide.com. 2015-03-25. Retrieved on 2015-10-05.
- "World Heritage sites participate in Earth Hour". UNESCO World Heritage Centre. March 28, 2012. Retrieved 2015-10-05.

5

Energy-efficient Landscaping: A Comprehensive Study

Energy efficient landscaping is designed to preserve energy. Some of the examples of energy efficient landscaping are planting of trees, earth sheltering and planting of windbreaks. This segment expounds topics such as low energy house, green building, passive house and passive cooling.

Energy-efficient Landscaping

Energy-efficient landscaping is a type of landscaping designed for the purpose of conserving energy. There is a distinction between the embedded energy of materials and constructing the landscape, and the energy consumed by the maintenance and operations of a landscape.

Design techniques include:

- Planting trees for the purpose of providing shade, which reduces cooling costs.
- Planting or building windbreaks to slow winds near buildings, which reduces heat loss.
- Wall sheltering, where shrubbery or vines are used to create a windbreak directly against a wall.
- Earth sheltering and positioning buildings to take advantage of natural landforms as windbreaks.
- Green roofs that cool buildings with extra thermal mass and evapotranspiration.
- Reducing the heat island effect with pervious paving, high albedo paving, shade, and minimizing paved areas.
- Site lighting with full cut off fixtures, light level sensors, and high efficiency fixtures

Energy-efficient landscaping techniques include using local materials, on-site composting and chipping to reduce greenwaste hauling, hand tools instead of gasoline-powered, and also may involve using drought-resistant plantings in arid areas, buying stock from local growers to avoid energy in transportation, and similar techniques.

Low-energy House

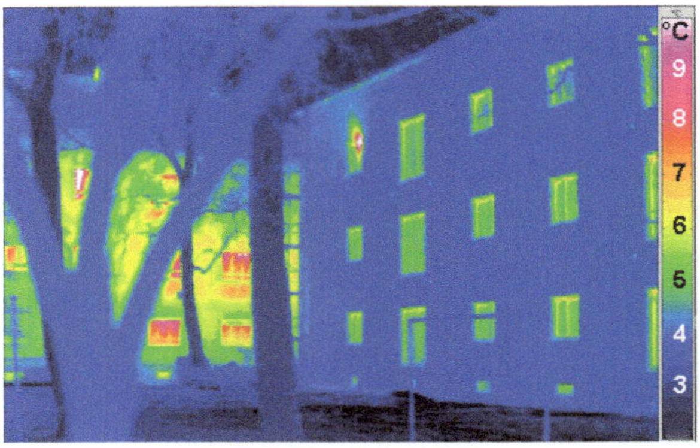

A thermogram compares the "heat radiation" of the windows and walls of two buildings: sustainable low-energy passive house (right) and conventional leaking house (left)

A low-energy house is any type of house that from design, technologies and building products uses less energy, from any source, than a traditional or average contemporary house. In the practice of sustainable design, sustainable architecture, low-energy building, energy-efficient landscaping low-energy houses often use active solar and passive solar building design techniques and components to reduce their energy expenditure.

General usage

The meaning of the term 'low-energy house' has changed over time, but in Europe it generally refers to a house that uses around half of the German or Swiss low-energy standards referred to below for space heating, typically in the range from 30 kWh/m^2a to 20 kWh/m^2a (9,500 Btu/ft^2/yr to 6,300 Btu/ft^2/yr). Below this the term 'Ultra-low-energy building' is often used.

The term can also refer to any dwelling whose energy use is below the standards demanded by current building codes. Because national standards vary considerably around the world, 'low-energy' developments in one country may not meet 'normal practice' in another.

National Standards

In some countries the term relates to a specific building standard. In particular, these seek to limit the energy used for space heating, since in many climate zones it represents the largest energy use. Other energy use may also be regulated. The history of passive solar building design gives an international look at one form of low-energy building development and standards.

Europe

In Germany a low-energy house *(Niedrigenergiehaus)* has a limit equivalent to 7 litres of heating oil for each square metre of room *for space heating* annually (50 kWh/m^{2}a or 15,850 Btu/ft^2/yr). In Switzerland, the term is used in connection with the *MINERGIE* standard (42 kWh/m^{2}a or 13,300 Btu/ft^2/yr) or the Minergie-P (equivalent to the Passivhaus).

In comparison, the German *Passivhaus* ultra-low-energy standard, currently undergoing adoption in some other European countries, has a maximum space heating requirement of 15 kWh/m^{2}a or 4,755 Btu/ft^2/yr.

A "sub-10 passive house" is under construction in Ireland that has an independently evaluated PHPP (Passive House) rating of 9.5 kW/m^2/year. Its form of construction also tackles the issue of embodied energy, which can significantly distort the lifecycle CO_2 emissions associated with even low energy use houses.

North America

In the United States, the ENERGY STAR program is the largest program defining low-energy homes and consumer products. Homes earning ENERGY STAR certification use at least 15% less energy than standard new homes built to the International Residential Code, although homes typically achieve 20%–30% savings.

In addition, the US Department of Energy launched a program in 2008 with the goal of spreading zero-energy housing over the US. Currently, participating builders commit to constructing new homes that achieve 30% savings on a home energy rating scale.

Zero-energy and Energy-plus Buildings

Beyond ultra-low-energy buildings are those that use, on average over the course of a year, no imported energy - zero-energy buildings – or even those that generate a surplus - energy-plus houses – both of which have been and are being successfully built.

This can be achieved by a mixture of energy conservation technologies and the use of renewable energy sources. However, in the absence of recognized standards, the mix between these – and consequently the energy-use profile and environmental impact of the building – can vary significantly.

At one end of the spectrum are buildings with an ultra-low space heating requirement that therefore require low levels of imported energy, even in winter, approaching the concept of an autonomous building.

At the opposite end of the spectrum are buildings where few attempts are made to reduce the space heating requirement and which therefore use high levels of imported energy in winter. While this can be balanced by high levels of renewable energy generation throughout the year, it imposes greater demands on the traditional national energy infrastructure during the peak winter season.

- Zero-energy building
- Superinsulation
- PlusEnergy

Low-energy Technology

Introduction

Low-energy buildings typically use high levels of insulation, energy efficient windows, low levels of air infiltration and heat recovery ventilation to lower heating and cooling energy. They may also use passive solar building design techniques or active solar technologies. These homes may use hot water heat recycling technologies to recover heat from showers and dishwashers. Lighting and miscellaneous energy use is alleviated with fluorescent lighting and efficient appliances. Weatherization provides more information on increasing building energy efficiency.

Passive Houses are required to achieve a whole building air change rate of no more than 0.6 ac/hr under forced pressurisation and depressurisation testing at 50Pa minimum. On site blower door testing by certified testers is used to prove compliance.

A significant feature of ultra-low-energy buildings is the increasing importance of heat loss through linear thermal bridging within the construction. Failure to eliminate thermal pathways from warm to cold surfaces ("bridges") creates the conditions for interstitial condensation forming deep within the construction and lead to potentially serious issues of mould growth and rot. With near zero filtration losses through the fabric of the dwelling, air movement cannot be relied upon to dry out the construction and a comprehensive condensation risk analysis of every abutment detail is recommended.

Improvements to Heating, Cooling, Ventilation and Water Heating

• Absorption refrigerator • Annualized geothermal solar • Earth cooling tubes • Geothermal heat pump • Heat recovery ventilation • Hot water heat recycling	• Passive cooling • Renewable heat • Seasonal thermal energy storage (STES) • Solar air conditioning • Solar hot water • Solar devices

Passive solar Design and Landscape

Passive solar building design and energy-efficient landscaping support the low-energy house in conservation and can integrate them into a neighborhood and environment. Following passive solar building techniques, where possible buildings are compact in shape to reduce their surface area, with principal windows oriented towards the equator - south in the northern hemisphere and north in the southern hemisphere - to maximize passive solar gain. However, the use of solar gain, especially in temperate climate regions, is secondary to minimizing the overall house energy requirements. On the other hand in hot climates temperatures excess heat can create uncomfortable indoor conditions. Passive alternatives to air conditioning systems such as temperature-dependent venting have been shown to be effective in regions with cooling needs. Other techniques to combat excessive solar heat gains include Brise soleils, trees, attached pergolas with vines, vertical gardens, green roofs among others.

Low-energy houses can be constructed from dense or lightweight materials, but some internal thermal mass is normally incorporated to reduce summer peak temperatures, maintain stable winter temperatures, and prevent possible overheating in spring or autumn before the higher sun angle "shades" mid-day wall exposure and window penetration. Exterior wall color, when the surface allows choice, for reflection or absorption insolation qualities depends on the predominant year-round ambient outdoor temperature. The use of deciduous trees and wall trellised or self attaching vines can assist in climates not at the temperature extremes.

- Sustainable landscaping
- Sustainable landscape architecture
- Sustainable gardening
- Rainwater harvesting
- Water conservation

Lighting and Electrical Appliances

To minimize the total primary energy consumption, the many passive and active daylighting techniques are the first daytime solution to employ. For low light level days, non-daylighted spaces, and nighttime; the use of creative-sustainable lighting design using low-energy sources such as 'standard voltage' compact fluorescent lamps and solid-state lighting with Light-emitting diode-LED lamps, organic light-emitting diodes, and PLED - polymer light-emitting diodes; and 'low voltage' electrical filament-Incandescent light bulbs, and compact Metal halide, Xenon and Halogen lamps, can be used.

Solar powered exterior circulation, security, and landscape lighting - with photovoltaic cells on each fixture or connecting to a central Solar panel system, are available for gardens and outdoor needs. Low voltage systems can be used for more controlled or independent illumination, while still using less electricity than conventional fixtures and lamps. Timers, motion detection and natural light operation sensors reduce energy consumption, and light pollution even further for a Low-energy house setting.

Appliance consumer products meeting independent energy efficiency testing and receiving Ecolabel certification marks for reduced electrical-'natural-gas' consumption and product manufacturing carbon emission labels are preferred for use in Low-energy houses. The ecolabel certification marks of Energy Star and EKOenergy are examples.

Green Building

US EPA Kansas City Science & Technology Center. This facility features the following green attributes:

- LEED 2.0 Gold certified
- Green Power
- Native Landscaping

Green building (also known as green construction or sustainable building) refers to both a structure and the using of processes that are environmentally responsible and resource-efficient throughout a building's life-cycle: from siting to design, construction, operation, maintenance, renovation, and demolition. In other words, green building design involves finding the balance between homebuilding and the sustainable environment. This requires close cooperation of the design team, the architects, the engineers, and the client at all project stages. The Green Building practice expands and complements the classical building design concerns of economy, utility, durability, and comfort.

Leadership in Energy and Environmental Design (LEED) is a set of rating systems for the design, construction, operation, and maintenance of green buildings which was Developed by the U.S. Green Building Council. Other certificates system that confirms the sustainability of buildings is the British BREEAM (Building Research Establishment Environmental Assessment Method) for buildings and large scale developments. Currently, World Green Building Council is conducting research on the effects of green buildings on the health and productivity of their users and is working with World Bank to promote Green Buildings in Emerging Markets through EDGE Excellence in Design for Greater Efficiencies Market Transformation Program and certification.

Although new technologies are constantly being developed to complement current practices in creating greener structures, the common objective of green buildings is to reduce the overall impact of the built environment on human health and the natural environment by:

- Efficiently using energy, water, and other resources
- Protecting occupant health and improving employee productivity
- Reducing waste, pollution and environmental degradation

A similar concept is natural building, which is usually on a smaller scale and tends to focus on the use of natural materials that are available locally. Other related topics include sustainable design and green architecture. Sustainability may be defined as meeting the needs of present generations without compromising the ability of future generations to meet their needs. Although some green building programs don't address the issue of the retrofitting existing homes, others do, especially through public schemes for energy efficient refurbishment. Green construction principles can easily be applied to retrofit work as well as new construction.

A 2009 report by the U.S. General Services Administration found 12 sustainably-designed buildings that cost less to operate and have excellent energy performance. In addition, occupants were overall more satisfied with the building than those in typical commercial buildings.These are eco-friendly buildings.

Reducing Environmental Impact

Hanging gardens of One Central Park, Sydney

Globally buildings are responsible for a huge share of energy, electricity, water and materials consumption. The building sector has the greatest potential to deliver significant cuts in emissions at little or no cost. Buildings account for 18% of global emissions today, or the equivalent of 9 billion tonnes of CO_2 annually. If new technologies in construction are not adopted during this time of rapid growth, emissions could double by 2050, according to the United Nations Environment Program.Green building practices aim to reduce the environmental impact of building. Since construction almost always degrades a building site, not building at all is preferable to green building, in terms of reducing environmental impact. The second rule is that every building should be as small as possible. The third rule is not to contribute to sprawl, even if the most energy-efficient, environmentally sound methods are used in design and construction.

Buildings account for a large amount of land. According to the National Resources Inventory, approximately 107 million acres (430,000 km^2) of land in the United States are developed. The International Energy Agency released a publication that estimated that existing buildings are responsible for more than 40% of the world's total primary energy consumption and for 24% of global carbon dioxide emissions.

Goals of Green Building

Blu Homes mkSolaire, a green building designed by Michelle Kaufmann.

Taipei 101, the tallest and largest green building of LEED Platinum certification in the world since 2011.

The concept of sustainable development can be traced to the energy (especially fossil oil) crisis and environmental pollution concerns of the 1960s and 1970s. The Rachel Carson book, "Silent Spring", published in 1962, is considered to be one of the first initial efforts to describe sustainable development as related to green building. The green building movement in the U.S. originated from the need and desire for more energy efficient and environmentally friendly construction practices. There are a number of motives for building green, including environmental, economic, and social benefits. However, modern sustainability initiatives call for an integrated and synergistic design to both new construction and in the retrofitting of existing structures. Also known as sustainable design, this approach integrates the building life-cycle with each green practice employed with a design-purpose to create a synergy among the practices used.

Green building brings together a vast array of practices, techniques, and skills to reduce and ultimately eliminate the impacts of buildings on the environment and human health. It often emphasizes taking advantage of renewable resources, e.g., using sunlight through passive solar, active solar, and photovoltaic equipment, and using plants and trees through green roofs, rain gardens, and reduction of rainwater run-off. Many other techniques are used, such as using low-impact building materials or using packed gravel or permeable concrete instead of conventional concrete or asphalt to enhance replenishment of ground water.

While the practices or technologies employed in green building are constantly evolving and may differ from region to region, fundamental principles persist from which the method is derived: siting and structure design efficiency, energy efficiency, water efficiency, materials efficiency, indoor environmental quality enhancement, operations and maintenance optimization and waste and toxics reduction. The essence of green building is an optimization of one or more of these principles. Also, with the proper synergistic design, individual green building technologies may work together to produce a greater cumulative effect.

On the aesthetic side of green architecture or sustainable design is the philosophy of designing a building that is in harmony with the natural features and resources surrounding the site. There are several key steps in designing sustainable buildings: specify 'green' building materials from local sources, reduce loads, optimize systems, and generate on-site renewable energy.

Life Cycle Assessment

A life cycle assessment (LCA) can help avoid a narrow outlook on environmental, social and economic concerns by assessing a full range of impacts associated with all cradle-to-grave stages of a process: from extraction of raw materials through materials processing, manufacture, distribution, use, repair and maintenance, and disposal or recycling. Impacts taken into account include (among others) embodied energy, global warming potential, resource use, air pollution, water pollution, and waste.

In terms of green building, the last few years have seen a shift away from a *prescriptive* approach, which assumes that certain prescribed practices are better for the environment, toward the scientific evaluation of actual performance through LCA.

Although LCA is widely recognized as the best way to evaluate the environmental impacts of buildings (ISO 14040 provides a recognized LCA methodology), it is not yet a consistent requirement of green building rating systems and codes, despite the fact that embodied energy and other life cycle impacts are critical to the design of environmentally responsible buildings.

In North America, LCA is rewarded to some extent in the Green Globes® rating system, and is part of the new American National Standard based on Green Globes, *ANSI/GBI 01-2010: Green Building Protocol for Commercial Buildings*. LCA is also included as a pilot credit in the LEED system, though a decision has not been made as to whether it will be incorporated fully into the next major revision. The state of California also included LCA as a voluntary measure in its 2010 draft *Green Building Standards Code*.

Although LCA is often perceived as overly complex and time consuming for regular use by design professionals, research organizations such as BRE in the UK and the Athena Sustainable Materials Institute in North America are working to make it more accessible.

In the UK, the BRE *Green Guide to Specifications* offers ratings for 1,500 building materials based on LCA.

In North America, the ATHENA® *EcoCalculator for Assemblies* provides LCA results for several hundred common building assembles based on data generated by its more complex parent software, the ATHENA® *Impact Estimator for Buildings*. (The EcoCalculator is available free at www.athenasmi.org.) Athena software tools are especially useful early in the design process when material choices have far-reaching implications for overall environmental impact. They allow designers to experiment with different material mixes to achieve the most effective combination.

Siting and Structure Design Efficiency

The foundation of any construction project is rooted in the concept and design stages. The concept stage, in fact, is one of the major steps in a project life cycle, as it has the largest impact on cost and performance. In designing environmentally optimal buildings, the objective is to minimize the total environmental impact associated with all life-cycle stages of the building project.

However, building as a process is not as streamlined as an industrial process, and varies from one building to the other, never repeating itself identically. In addition, buildings are much more complex products, composed of a multitude of materials and components each constituting various design variables to be decided at the design stage. A variation of every design variable may affect the environment during all the building's relevant life-cycle stages.

Exterior Light Shelves - Green Office Building, Denver Colorado

Energy Efficiency

An eco-house at Findhorn Ecovillage with a turf roof and solar panels

Green buildings often include measures to reduce energy consumption – both the embodied energy required to extract, process, transport and install building materials and operating energy to provide services such as heating and power for equipment.

As high-performance buildings use less operating energy, embodied energy has assumed much greater importance – and may make up as much as 30% of the overall life cycle energy consumption. Studies such as the U.S. LCI Database Project show buildings built primarily with wood will have a lower embodied energy than those built primarily with brick, concrete, or steel.

To reduce operating energy use, designers use details that reduce air leakage through the building envelope (the barrier between conditioned and unconditioned space). They also specify high-performance windows and extra insulation in walls, ceilings, and floors. Another strategy, passive solar building design, is often implemented in low-energy homes. Designers orient windows and walls and place awnings, porches, and trees to shade windows and roofs during the summer while maximizing solar gain in the winter. In addition, effective window placement (daylighting) can provide more natural light and lessen the need for electric lighting during the day. Solar water heating further reduces energy costs.

Onsite generation of renewable energy through solar power, wind power, hydro power, or biomass can significantly reduce the environmental impact of the building. Power generation is generally the most expensive feature to add to a building.

Water Efficiency

Reducing water consumption and protecting water quality are key objectives in sustainable building. One critical issue of water consumption is that in many areas, the demands on the supplying aquifer exceed its ability to replenish itself. To the maximum extent feasible, facilities should increase their dependence on water that is collected, used, purified, and reused on-site. The protection and conservation of water throughout the life of a building may be accomplished by designing for dual plumbing that recycles water in toilet flushing or by using water for washing of the cars. Waste-water may be minimized by utilizing water conserving fixtures such as ultra-low flush toilets and low-flow shower heads. Bidets help eliminate the use of toilet paper, reducing sewer traffic and increasing possibilities of re-using water on-site. Point of use water treatment and heating improves both water quality and energy efficiency while reducing the amount of water in circulation. The use of non-sewage and greywater for on-site use such as site-irrigation will minimize demands on the local aquifer.

Large commercial buildings with water and energy efficiency can qualify for an LEED Certification. Philadelphia's Comcast Center is the tallest building in Philadelphia. It's also one of the tallest buildings in the USA that is LEED Certified. Their environmental engineering consists of a hybrid central chilled water system which cools floor-by-floor with steam instead of water. Burn's Mechanical set-up the entire renovation of the 58 story, 1.4 million square foot sky scraper.

Materials Efficiency

Building materials typically considered to be 'green' include lumber from forests that have been certified to a third-party forest standard, rapidly renewable plant materials like bamboo and straw, dimension stone, recycled stone, recycled metal, and other products that are non-toxic, reusable, re-newable, and/or recyclable. For concrete a high performance or Roman self-healing concrete is available. The EPA (Environmental Protection Agency) also suggests using recycled industrial goods, such as coal combustion products, foundry sand, and demolition debris in construction projects. Energy efficient building materials and appliances are promoted in the United States through energy rebate programs.

Indoor Environmental Quality Enhancement

The Indoor Environmental Quality (IEQ) category in LEED standards, one of the five environmental categories, was created to provide comfort, well-being, and productivity of occupants. The LEED IEQ category addresses design and construction guidelines especially: indoor air quality (IAQ), thermal quality, and lighting quality.

Indoor Air Quality seeks to reduce volatile organic compounds, or VOCs, and other air impurities such as microbial contaminants. Buildings rely on a properly designed ventilation system (passively/naturally or mechanically powered) to provide adequate ventilation of cleaner air from outdoors or recirculated, filtered air as well as isolated operations (kitchens, dry cleaners, etc.) from other occupancies. During the design and construction process choosing construction materials and interior finish products with zero or low VOC emissions will improve IAQ. Most building materials and cleaning/maintenance products emit gases, some of them toxic, such as many VOCs including formaldehyde. These gases can have a detrimental impact on occupants' health, comfort, and productivity. Avoiding these products will increase a building's IEQ. LEED, HQE and Green Star contain specifications on use of low-emitting interior. Draft LEED 2012 is about to expand the scope of the involved products. BREEAM limits formaldehyde emissions, no other VOCs. MAS Certified Green is a registered trademark to delineate low VOC-emitting products in the marketplace. The MAS Certified Green Program ensures that any potentially hazardous chemicals released from manufactured products have been thoroughly tested and meet rigorous standards established by independent toxicologists to address recognized long term health concerns. These IAQ standards have been adopted by and incorporated into the following programs: (1) The United States Green Building Council (USGBC) in their LEED rating system (2) The California Department of Public Health (CDPH) in their section 01350 standards (3) The Collaborative for High Performance Schools (CHPS) in their Best Practices Manual and (4) The Business and Institutional Furniture Manufacturers Association (BIFMA) in their level® sustainability standard.

Also important to indoor air quality is the control of moisture accumulation (dampness) leading to mold growth and the presence of bacteria and viruses as well as dust mites and other organisms and microbiological concerns. Water intrusion through a building's envelope or water condensing on cold surfaces on the building's interior can enhance and sustain microbial growth. A well-insulated and tightly sealed envelope will reduce moisture problems but adequate ventilation is also necessary to eliminate moisture from sources indoors including human metabolic processes, cooking, bathing, cleaning, and other activities.

Personal temperature and airflow control over the HVAC system coupled with a properly designed building envelope will also aid in increasing a building's thermal quality. Creating a high performance luminous environment through the careful integration of daylight and electrical light sources will improve on the lighting quality and energy performance of a structure.

Solid wood products, particularly flooring, are often specified in environments where occupants are known to have allergies to dust or other particulates. Wood itself is considered to be hypo-allergenic and its smooth surfaces prevent the buildup of particles common in soft finishes like carpet. The Asthma and Allergy Foundation of American recommends hardwood, vinyl, linoleum tile or slate flooring instead of carpet. The use of wood products can also improve air quality by absorbing or releasing moisture in the air to moderate humidity.

Interactions among all the indoor components and the occupants together form the processes that determine the indoor air quality. Extensive investigation of such processes is the subject of indoor air scientific research and is well documented in the journal Indoor Air.

Operations and Maintenance Optimization

No matter how sustainable a building may have been in its design and construction, it can only remain so if it is operated responsibly and maintained properly. Ensuring operations and maintenance(O&M) personnel are part of the project's planning and development process will help retain the green criteria designed at the onset of the project. Every aspect of green building is integrated into the O&M phase of a building's life. The addition of new green technologies also falls on the O&M staff. Although the goal of waste reduction may be applied during the design, construction and demolition phases of a building's life-cycle, it is in the O&M phase that green practices such as recycling and air quality enhancement take place. O&M staff should aim to establish best practices in energy efficiency, resource conservation, ecologically sensitive products and other sustainable practices. Education of building operators and occupants is key to effective implementation of sustainable strategies in O&M services.

Waste Reduction

Green architecture also seeks to reduce waste of energy, water and materials used during construction. For example, in California nearly 60% of the state's waste comes from commercial buildings During the construction phase, one goal should be to reduce the amount of material going to landfills. Well-designed buildings also help reduce the amount of waste generated by the occupants as well, by providing on-site solutions such as compost bins to reduce matter going to landfills.

To reduce the amount of wood that goes to landfill, Neutral Alliance (a coalition of government, NGOs and the forest industry) created the website dontwastewood.com. The site includes a variety of resources for regulators, municipalities, developers, contractors, owner/operators and individuals/homeowners looking for information on wood recycling.

When buildings reach the end of their useful life, they are typically demolished and hauled to landfills. Deconstruction is a method of harvesting what is commonly considered "waste" and reclaiming it into useful building material. Extending the useful life of a structure also reduces waste – building materials such as wood that are light and easy to work with make renovations easier.

To reduce the impact on wells or water treatment plants, several options exist. "Greywater", wastewater from sources such as dishwashing or washing machines, can be used for subsurface irrigation, or if treated, for non-potable purposes, e.g., to flush toilets and wash cars. Rainwater collectors are used for similar purposes.

Centralized wastewater treatment systems can be costly and use a lot of energy. An alternative to this process is converting waste and wastewater into fertilizer, which avoids these costs and shows other benefits. By collecting human waste at the source and running it to a semi-centralized biogas plant with other biological waste, liquid fertilizer can be produced. This concept was demonstrated by a settlement in Lubeck Germany in the late 1990s. Practices like these provide soil with organic nutrients and create carbon sinks that remove carbon dioxide from the atmosphere, offsetting greenhouse gas emission. Producing artificial fertilizer is also more costly in energy than this process.

Cost and Payoff

The most criticized issue about constructing environmentally friendly buildings is the price. Photo-voltaics, new appliances, and modern technologies tend to cost more money. Most green buildings cost a premium of <2%, but yield 10 times as much over the entire life of the building. In regards to the financial benefits of green building, "Over 20 years, the financial payback typically exceeds the additional cost of greening by a factor of 4-6 times. And broader benefits, such as reductions in greenhouse gases (GHGs) and other pollutants have large positive impacts on surrounding communities and on the planet." The stigma is between the knowledge of up-front cost vs. life-cycle cost. The savings in money come from more efficient use of utilities which result in decreased energy bills. It is projected that different sectors could save $130 Billion on energy bills. Also, higher worker or student productivity can be factored into savings and cost deductions.

Numerous studies have shown the measurable benefit of green building initiatives on worker productivity. In general it has been found that, "there is a direct correlation between increased productivity and employees who love being in their work space." Specifically, worker productivity can be significantly impacted by certain aspects of green building design such as improved lighting, reduction of pollutants, advanced ventilation systems and the use of non-toxic building materials. In "The Business Case for Green Building", the U.S. Green Building Council gives another specific example of how commercial energy retrofits increase worker health and thus productivity, "People in the U.S. spend about 90% of their time indoors. EPA studies indicate indoor levels of pollutants may be up to ten times higher than outdoor levels. LEED-certified buildings are designed to have healthier, cleaner indoor environmental quality, which means health benefits for occupants."

Studies have shown over a 20-year life period, some green buildings have yielded $53 to $71 per square foot back on investment. Confirming the rentability of green building investments, further studies of the commercial real estate market have found that LEED and Energy Star certified buildings achieve significantly higher rents, sale prices and occupancy rates as well as lower capitalization rates potentially reflecting lower investment risk.

Regulation and Operation

As a result of the increased interest in green building concepts and practices, a number of organizations have developed standards, codes and rating systems that let government regulators, building professionals and consumers embrace green building with confidence. In some cases, codes are written so local governments can adopt them as bylaws to reduce the local environmental impact of buildings.

Green building rating systems such as BREEAM (United Kingdom), LEED (United States and Canada), DGNB (Germany), CASBEE (Japan), and VERDE^{GBCe} (Spain) help consumers determine a structure's level of environmental performance. They award credits for optional building features that support green design in categories such as location and maintenance of building site, conservation of water, energy, and building materials, and occupant comfort and health. The number of credits generally determines the level of achievement.

Green building codes and standards, such as the International Code Council's draft International Green Construction Code, are sets of rules created by standards development organizations that establish minimum requirements for elements of green building such as materials or heating and cooling.

Some of the major building environmental assessment tools currently in use include:

- United States: International Green Construction Code (IGCC)

International Frameworks and Assessment Tools

IPCC Fourth Assessment Report

Climate Change 2007, the Fourth Assessment Report (AR4) of the United Nations Intergovernmental Panel on Climate Change (IPCC), is the fourth in a series of such reports. The IPCC was established by the World Meteorological Organization (WMO) and the United Nations Environment Programme (UNEP) to assess scientific, technical and socio-economic information concerning climate change, its potential effects and options for adaptation and mitigation.

UNEP and Climate change

United Nations Environment Program UNEP works to facilitate the transition to low-carbon societies, support climate proofing efforts, improve understanding of climate change science, and raise public awareness about this global challenge.

GHG Indicator

The Greenhouse Gas Indicator: UNEP Guidelines for Calculating Greenhouse Gas Emissions for Businesses and Non-Commercial Organizations

Agenda 21

Agenda 21 is a programme run by the United Nations (UN) related to sustainable development. It is a comprehensive blueprint of action to be taken globally, nationally and locally by organizations of the UN, governments, and major groups in every area in which humans impact on the environment. The number 21 refers to the 21st century.

FIDIC's PSM

The International Federation of Consulting Engineers (FIDIC) Project Sustainability Management Guidelines were created in order to assist project engineers and other stakeholders in setting sustainable development goals for their projects that are recognized and accepted by as being in the interests of society as a whole. The process is also intended to allow the alignment of project goals with local conditions and priorities and to assist those involved in managing projects to measure and verify their progress.

The Project Sustainability Management Guidelines are structured with Themes and Sub-Themes under the three main sustainability headings of Social, Environmental and Economic. For each individual Sub-Theme a core project indicator is defined along with guidance as to the relevance of that issue in the context of an individual project.

The Sustainability Reporting Framework provides guidance for organizations to use as the basis for disclosure about their sustainability performance, and also provides stakeholders a universally applicable, comparable framework in which to understand disclosed information.

The Reporting Framework contains the core product of the Sustainability Reporting Guidelines, as well as Protocols and Sector Supplements. The Guidelines are used as the basis for all reporting. They are the foundation upon which all other reporting guidance is based, and outline core content for reporting that is broadly relevant to all organizations regardless of size, sector, or location. The Guidelines contain principles and guidance as well as standard disclosures – including indicators – to outline a disclosure framework that organizations can voluntarily, flexibly, and incrementally, adopt.

Protocols underpin each indicator in the Guidelines and include definitions for key terms in the indicator, compilation methodologies, intended scope of the indicator, and other technical references.

Sector Supplements respond to the limits of a one-size-fits-all approach. Sector Supplements complement the use of the core Guidelines by capturing the unique set of sustainability issues faced by different sectors such as mining, automotive, banking, public agencies and others.

IPD Environment Code

The IPD Environment Code was launched in February 2008. The Code is intended as a good practice global standard for measuring the environmental performance of corporate buildings. Its aim is to accurately measure and manage the environmental impacts of corporate buildings and enable property executives to generate high quality, comparable performance information about their buildings anywhere in the world. The Code covers a wide range of building types (from offices to airports) and aims to inform and support the following;

- Creating an environmental strategy
- Inputting to real estate strategy
- Communicating a commitment to environmental improvement
- Creating performance targets
- Environmental improvement plans
- Performance assessment and measurement
- Life cycle assessments
- Acquisition and disposal of buildings
- Supplier management
- Information systems and data population
- Compliance with regulations
- Team and personal objectives

IPD estimate that it will take approximately three years to gather significant data to develop a robust set of baseline data that could be used across a typical corporate estate.

ISO 21931

ISO/TS 21931:2006, Sustainability in building construction—Framework for methods of assessment for environmental performance of construction works—Part 1: Buildings, is intended to provide a general framework for improving the quality and comparability of methods for assessing the environmental performance of buildings. It identifies and describes issues to be taken into account when using methods for the assessment of environmental performance for new or existing building properties in the design, construction, operation, refurbishment and deconstruction stages. It is not an assessment system in itself but is intended be used in conjunction with, and following the principles set out in, the ISO 14000 series of standards.

Passive House

A building based on the passive house concept in Darmstadt, Germany

Passive house (German: *Passivhaus*) is a rigorous, voluntary standard for energy efficiency in a building, reducing its ecological footprint. It results in ultra-low energy buildings that require little energy for space heating or cooling. A similar standard, *MINERGIE-P*, is used in Switzerland. The standard is not confined to residential properties; several office buildings, schools, kindergartens and a supermarket have also been constructed to the standard. Passive design is not an attachment or supplement to architectural design, but a design process that is integrated with architectural design. Although it is mostly applied to new buildings, it has also been used for refurbishments.

Estimates of the number of Passivhaus buildings around the world in late 2008 ranged from 15,000 to 20,000 structures. As of August 2010, there were approximately 25,000 such certified structures of all types in Europe, while in the United States there were only 13, with a few dozen more under construction. The vast majority of passive structures have been built in German-speaking countries and Scandinavia.

History

Bo Adamson, co-originator of the passive house concept

Wolfgang Feist, co-originator of the passive house concept, and founder of the Passivhaus Institut in Germany

The Passivhaus standard originated from a conversation in May 1988 between Bo Adamson of Lund University, Sweden, and Wolfgang Feist of the Institut für Wohnen und Umwelt (Institute for Housing and the Environment, Germany). Their concept was developed through a number of research projects, aided by financial assistance from the German state of Hessen.

First Examples

The eventual building of four row houses (terraced houses or town homes) was designed for four private clients by the architectural firm of Bott, Ridder and Westermeyer. The first Passivhaus residences were built in Darmstadt, Germany in 1990, and occupied by the clients the following year.

Further Implementation and Councils

In September 1996 the Passivhaus-Institut was founded, also in Darmstadt, to promote and control the standards. Since then, thousands of Passivhaus structures have been built, to an estimated 25,000+ as of 2010. Most are located in Germany and Austria, with others in various countries worldwide.

After the concept had been validated at Darmstadt, with space heating 90% less than required for a standard new building of the time, the Economical Passive Houses Working Group was created in 1996. This group developed the planning package and initiated the production of the innovative components that had been used, notably the windows and the high-efficiency ventilation systems. Meanwhile, further passive houses were built in Stuttgart (1993), Naumburg, Hesse, Wiesbaden, and Cologne (1997).

The products developed for the Passivhaus standard were further commercialised during and following the European Union sponsored CEPHEUS project, which proved the concept in five European countries over the winter of 2000–2001. In North America the first Passivhaus was built in Urbana, Illinois in 2003, and the first to be certified was built in 2006 near Bemidji, Minnesota in Camp Waldsee of the German Concordia Language Villages. The first US passive retrofit project was certified in July 2010: the remodeled 2,400 sf craftsman O'Neill house in Sonoma, California.

Ireland's first Passive House was built in 2005 by Tomas O'Leary, a Passive house designer and teacher. The house was called 'Out of the Blue'. Upon completion, Tomas moved into the building.

The world's first standardised passive prefabricated house was built in Ireland in 2005 by Scandinavian Homes, a Swedish company that has since built more passive houses in England and Poland.

Present Day

Estimates in 2008 of the number of passive houses around the world ranged from 15,000 to 20,000. The vast majority have been built in German-speaking countries or Scandinavia. The first certified passive house in the Antwerpen region of Belgium was built in 2010. In 2011 the city of Heidelberg in Germany initiated the Bahnstadt project, which was seen as the world's largest passive house building areas. A company in Qatar is planning the country's first passivhaus in 2013, the first in the region.

Standards

While some techniques and technologies were specifically developed for the Passive House standard, others, such as superinsulation, already existed, and the concept of passive solar building design dates back to antiquity. There was also other previous experience with low-energy building standards, notably the German *Niedrigenergiehaus* (low-energy house) standard, as well as from buildings constructed to the demanding energy codes of Sweden and Denmark.

The Passivhaus standard requires that the building fulfills the following requirements:

- The building must be designed to have an annual heating and cooling demand as calculated with the Passivhaus Planning Package of not more than 15 kWh/m^2 (4,755 BTU/sq ft; 5.017 MJ/sq ft) per year in heating or cooling energy OR be designed with a peak heat load of 10 W/m^2 (1.2 hp/1000 sq ft)
- Total primary energy (source energy for electricity, etc.) consumption (primary energy for heating, hot water and electricity) must not be more than 120 kWh/m^2 (38,040 BTU/sq ft; 40.13 MJ/sq ft) per year
- The building must not leak more air than 0.6 times the house volume per hour ($n_{50} \leq 0.6$ / hour) at 50 Pa (0.0073 psi) as tested by a blower door, or alternatively when looked at the surface area of the enclosure, the leakage rate must be less than 0.05 cubic feet per minute

Recommendations

- Further, the specific heat load for the heating source at design temperature is recommended, but not required, to be less than 10 W/m^2 (3.17 btu/h.ft^2).

These standards are much higher than houses built to most normal building codes.

National partners within the 'consortium for the Promotion of European Passive Houses' are thought to have some flexibility to adapt these limits locally.

Space Heating Requirement

By achieving the Passivhaus standards, qualified buildings are able to dispense with conventional heating systems. While this is an underlying objective of the Passivhaus standard, some type of heating will still be required and most Passivhaus buildings do include a system to provide supplemental space heating. This is normally distributed through the low-volume heat recovery ventilation system that is required to maintain air quality, rather than by a conventional hydronic or high-volume forced-air heating system, as described in the space heating section below.

Construction Costs

In Passivhaus buildings, the cost savings from dispensing with the conventional heating system can be used to fund the upgrade of the building envelope and the heat recovery ventilation system. With careful design and increasing competition in the supply of the specifically designed Passivhaus building products, in Germany it is now possible to construct buildings for the same cost as those built to normal German building standards, as was done with the Passivhaus apartments at Vauban, Freiburg. On average passive houses are reported to be more expensive upfront than conventional buildings - 5% to 8% in Germany, 8% to 10% in UK and 5% to 10% in USA.

Evaluations have indicated that while it is technically possible, the costs of meeting the Passivhaus standard increase significantly when building in Northern Europe above 60° latitude. European cities at approximately 60° include Helsinki in Finland and Bergen in Norway. London is at 51°; Moscow is at 55°.

These facts have led a number of architects to construct buildings that use the ground under the building for massive heat storage to shift heat production from the winter to the summer. Some buildings can also shift cooling from the summer to the winter. At least one designer uses a passive thermosiphon carrying only air, so the process can be accomplished without expensive, unreliable machinery.

Design and Construction

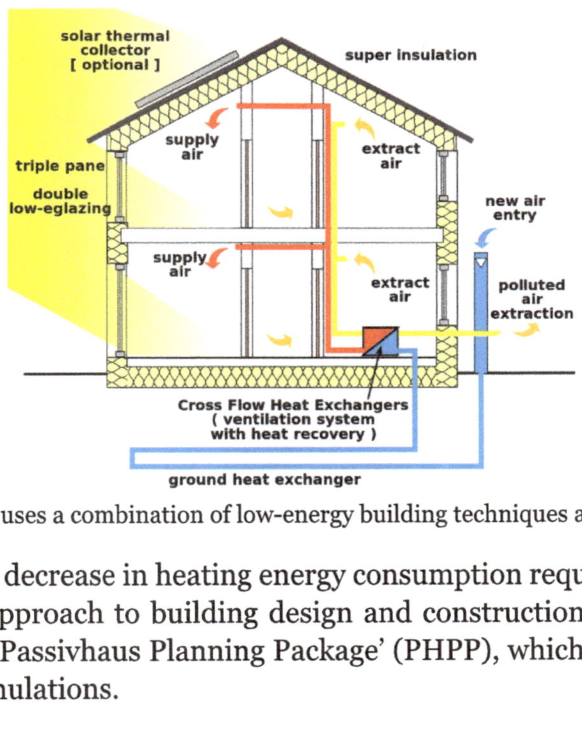

The Passivhaus uses a combination of low-energy building techniques and technologies

Achieving the major decrease in heating energy consumption required by the standard involves a shift in approach to building design and construction. Design may be assisted by use of the 'Passivhaus Planning Package' (PHPP), which uses specifically designed computer simulations.

To achieve the standards, a number of techniques and technologies are used in combination:

Passive solar Design and Landscape

Passive solar building design and energy-efficient landscaping support the Passive house energy conservation and can integrate them into a neighborhood and environment. Following passive solar building techniques, where possible buildings are compact in shape to reduce their surface area, with principal windows oriented towards the equator - south in the northern hemisphere and north in the southern hemisphere - to maximize passive solar gain. However, the use of solar gain, especially in temperate climate regions, is secondary to minimizing the overall house energy requirements. In climates and regions needing to reduce excessive summer passive solar heat gain, whether from direct or reflected sources, *Brise soleil*, trees, attached *pergolas* with vines, vertical gardens, green roofs, and other techniques are implemented.

Passive houses can be constructed from dense or lightweight materials, but some internal thermal mass is normally incorporated to reduce summer peak temperatures, maintain stable winter temperatures, and prevent possible overheating in spring or autumn before the higher sun angle "shades" mid-day wall exposure and window penetration. Exterior wall color, when the surface allows choice, for reflection or absorption insolation qualities depends on the predominant year-round ambient outdoor temperature. The use of deciduous trees and wall trellised or self attaching vines can assist in climates not at the temperature extremes.

Superinsulation

Passivhaus buildings employ superinsulation to significantly reduce the heat transfer through the walls, roof and floor compared to conventional buildings. A wide range of thermal insulation materials can be used to provide the required high R-values (low U-values, typically in the 0.10 to 0.15 $W/(m^2.K)$ range). Special attention is given to eliminating thermal bridges.

A disadvantage resulting from the thickness of wall insulation required is that, unless the external dimensions of the building can be enlarged to compensate, the internal floor area of the building may be less compared to traditional construction.

In Sweden, to achieve passive house standards, the insulation thickness would be 335 mm (about 13 in) (0.10 $W/(m^2.K)$) and the roof 500 mm (about 20 in) (U-value 0.066 $W/(m^2.K)$).

Advanced Window Technology

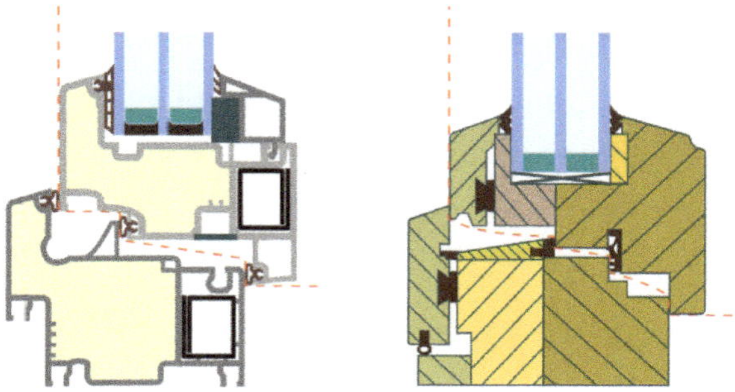

Typical Passive House windows

To meet the requirements of the Passivhaus standard, windows are manufactured with exceptionally high R-values (low U-values, typically 0.85 to 0.70 W/(m^2.K) for the entire window including the frame). These normally combine triple-pane insulated glazing (with a good solar heat-gain coefficient, low-emissivity coatings, sealed argon or krypton gas filled inter-pane voids, and 'warm edge' insulating glass spacers) with air-seals and specially developed thermally broken window frames.

In Central Europe and most of the United States, for unobstructed south-facing Passivhaus windows, the heat gains from the sun are, on average, greater than the heat losses, even in mid-winter.

Airtightness

Building envelopes under the Passivhaus standard are required to be extremely airtight compared to conventional construction. They are required to meet either 0.60 ACH50 (air changes per hour at 50 pascals) based on the building's volume, or 0.05 CFM50/sf (cubic feet per minute at 50 pascals, per square foot of building enclosure surface area). In order to achieve these metrics, recommended best practice is to test the building air barrier enclosure with a blower door at mid-construction if possible.

Passive house is designed so that most of the air exchange with exterior is done by controlled ventilation through a heat-exchanger in order to minimize heat loss (or gain, depending on climate), so uncontrolled air leaks are best avoided. Another reason is the passive house standard makes extensive use of insulation which usually requires a careful management of moisture and dew points. This is achieved through air barriers, careful sealing of every construction joint in the building envelope, and sealing of all service penetrations.

Ventilation

Use of passive natural ventilation is an integral component of passive house design where ambient temperature is conducive — either by singular or cross ventilation, by a simple opening or enhanced by the stack effect from smaller ingress with larger egress windows and/or clerestory-operable skylight.

When ambient climate is not conducive, mechanical heat recovery ventilation systems, with a heat recovery rate of over 80% and high-efficiency electronically commutated motors (ECM), are employed to maintain air quality, and to recover sufficient heat to dispense with a conventional central heating system. Since passively designed buildings are essentially air-tight, the rate of air change can be optimized and carefully controlled at about 0.4 air changes per hour. All ventilation ducts are insulated and sealed against leakage.

Some Passivhaus builders promote the use of earth warming tubes (typically ≈200 mm (~7,9 in) diameter, ≈40 m (~130 ft) long at a depth of ≈1.5 m (~5 ft)). These are buried in the soil to act as earth-to-air heat exchangers and pre-heat (or pre-cool) the intake air for the ventilation system. In cold weather the warmed air also prevents ice formation in the heat recovery system's heat exchanger. Concerns about this technique have arisen in some climates due to problems with condensation and mold.

Alternatively, an earth to air heat exchanger can use a liquid circuit instead of an air circuit, with a heat exchanger (battery) on the supply air.

Space Heating

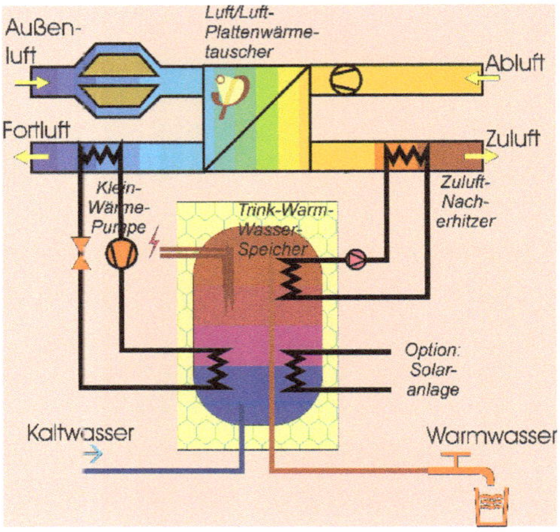

Passivhaus: In addition to the heat exchanger (centre), a micro-heat pump extracts heat from the exhaust air (left) and hot water heats the ventilation air (right). The ability to control building temperature using only the normal volume of ventilation air is fundamental.

In addition to using passive solar gain, Passivhaus buildings make extensive use of their intrinsic heat from internal sources—such as waste heat from lighting, white goods (major appliances) and other electrical devices (but not dedicated heaters)—as well as body heat from the people and other animals inside the building. This is due to the fact that people, on average, emit heat equivalent to 100 watts each of radiated thermal energy.

Together with the comprehensive energy conservation measures taken, this means that a conventional central heating system is not necessary, although they are sometimes installed due to client skepticism.

Instead, Passive houses sometimes have a dual purpose 800 to 1,500 watt heating and/or cooling element integrated with the supply air duct of the ventilation system, for use during the coldest days. It is fundamental to the design that all the heat required can be transported by the normal low air volume required for ventilation. A maximum air temperature of 50 °C (122 °F) is applied, to prevent any possible smell of scorching from dust that escapes the filters in the system.

The air-heating element can be heated by a small heat pump, by direct solar thermal energy, annualized geothermal solar, or simply by a natural gas or oil burner. In some cases a micro-heat pump is used to extract additional heat from the exhaust ventilation air, using it to heat either the incoming air or the hot water storage tank. Small wood-burning stoves can also be used to heat the water tank, although care is required to ensure that the room in which stove is located does not overheat.

Beyond the recovery of heat by the heat recovery ventilation unit, a well designed Passive house in the European climate should not need any supplemental heat source if the heating load is kept under 10W/m^2.

Because the heating capacity and the heating energy required by a passive house both are very low, the particular energy source selected has fewer financial implications than in a traditional building, although renewable energy sources are well suited to such low loads.

The Passive House Standards in Europe determine a Space Heating and cooling Energy Demand of 15 kilowatt hours per square meter of Treated Floor Area per year or 10 Watts per square meter peak demand. (Or in Imperial units 4.75 kBTU/sf*yr and 3.2 BTU/hr*sf respectively.) In addition, the total energy to be used in the building operations including heating, cooling, lighting, equipment, hot water, plug loads, etc. is limited to 120 kilowatt hours per square meter of Treated Floor Area per year. (Or in Imperial units 38.0 BTU/sf*yr.)

Lighting and Electrical Appliances

To minimize the total primary energy consumption, the many passive and active daylighting techniques are the first daytime solution to employ. For low light level days, non-daylighted spaces, and nighttime; the use of creative-sustainable lighting design using low-energy sources such as 'standard voltage' compact fluorescent lamps and solid-state lighting with Light-emitting diode-LED lamps, organic light-emitting diodes, and PLED - polymer light-emitting diodes; and 'low voltage' electrical filament-Incandescent light bulbs, and compact Metal halide, Xenon and Halogen lamps, can be used.

Solar powered exterior circulation, security, and landscape lighting - with photovoltaic cells on each fixture or connecting to a central Solar panel system, are available for gardens and outdoor needs. Low voltage systems can be used for more controlled or independent illumination, while still using less electricity than conventional fixtures and lamps. Timers, motion detection and natural light operation sensors reduce energy consumption, and light pollution even further for a Passivhaus setting.

Appliance consumer products meeting independent energy efficiency testing and receiving Ecolabel certification marks for reduced electrical-'natural-gas' consumption and product manufacturing carbon emission labels are preferred for use in Passive houses. The ecolabel certification marks of Energy Star and EKOenergy are examples.

Traits of Passive Houses

Typically, passive houses feature:

- Fresh, clean air: Note that for the parameters tested, and provided the filters (minimum F6) are maintained, HEPA quality air is provided. 0.3 air changes per hour (ACH) are recommended, otherwise the air can become "stale" (excess CO_2, flushing of indoor air pollutants) and any greater, excessively dry (less than 40% humidity). This implies careful selection of interior finishes and furnishings, to minimize indoor air pollution from VOC's (e.g., formaldehyde). This can be counteracted somewhat by opening a window for a very brief time, by plants, and by indoor fountains.
- Because of the high resistance to heat flow (high R-value insulation), there are no "outside walls" which are colder than other walls.
- Homogeneous interior temperature: it is impossible to have single rooms (e.g. the sleeping rooms) at a different temperature from the rest of the house. Note that the relatively high temperature of the sleeping areas is physiologically not considered desirable by some building scientists. Bedroom windows can be cracked open slightly to alleviate this when necessary.
- Slow temperature changes: with ventilation and heating systems switched off, a passive house typically loses less than 0.5 °C (1 °F) per day (in winter), stabilizing at around 15 °C (59 °F) in the central European climate.
- Quick return to normal temperature: opening windows or doors for a short time has only a limited effect; after apertures are closed, the air very quickly returns to the "normal" temperature.
- Some have voiced concerns that Passivhaus is not a general approach as the occupant has to behave in a prescribed way, for example not opening windows too often. However modelling shows that such concerns are not valid.

International Comparisons

- In the United States, a house built to the Passive House standard results in a building that requires space heating energy of 1 BTU per square foot (11 kJ/m^2) per heating degree day, compared with about 5 to 15 BTUs per square foot (56-170 kJ/m^2) per heating degree day for a similar building built to meet the 2003 Model Energy Efficiency Code. This is between 75 and 95% less energy for space heating and cooling than current new buildings that meet today's US energy efficiency codes. The Passivhaus in the German-language camp of Waldsee, Minnesota was designed under the guidance of architect Stephan Tanner of INTEP, LLC, a Minneapolis- and Munich-based consulting company for high performance and sustainable construction. Waldsee BioHaus is modeled on Germany's Passivhaus standard: beyond that of the U.S. LEED standard which improves quality of life inside the building while using 85% less energy than a house built to Minnesota building codes. VOLKsHouse 1.0 was the first certified Passive House offered and sold in Santa Fe New Mexico.

- In the United Kingdom, an average new house built to the Passive House standard would use 77% less energy for space heating, compared to the circa-2006 Building Regulations.
- In Ireland, it is calculated that a typical house built to the Passive House standard instead of the 2002 Building Regulations would consume 85% less energy for space heating and cut space-heating related carbon emissions by 94%.

Comparison With Zero Energy Buildings

A net zero-energy building (ZEB) is a building that over a year does not use more energy than it generates. The first 1979 Zero Energy Design building used passive solar heating and cooling techniques with air-tight construction and super insulation. A few ZEB's fail to fully exploit more affordable conservation technology and all use onsite active renewable energy technologies like photovoltaic to offset the building's primary energy consumption. Passive House and ZEB are complementary synergistic technology approaches, based on the same physics of thermal energy transfer and storage: ZEBs drive the annual energy consumption down to 0 kWh/m^2 with help from on-site renewable energy sources and can benefit from materials and methods which are used to meet the Passive House demand constraint of 120 kWh/m^2 which will minimize the need for the often costly on-site renewable energy sources. Energy Plus houses are similar to both PassivHaus and ZEB but emphasize the production of more energy per year than they consume, e.g., annual energy performance of -25 kWh/m^2 is an Energy Plus house.

Tropical Climate Needs

In a tropical climate, it could be helpful for ideal internal conditions to use Energy Recovery Ventilation instead of Heat Recovery Ventilation to reduce the humidity load of ventilation on the mechanical dehumidification system. Although dehumidifiers might be used, heat pump hot water heaters also will act to cool and condense interior humidity (where it can be dumped into drains) and dump the heat into the hot water tank. Passive cooling, solar air conditioning, and other solutions in passive solar building design need to be studied to adapt the Passive house concept for use in more regions of the world.

There is a certified Passive House in the hot and humid climate of Lafayette, Louisiana, USA, which uses Energy Recovery Ventilation and an efficient one ton air-conditioner to provide cooling and dehumidification.

Solar access is a very important factor in any design of a passive house as it allows the structure to use the solar energy to heat and light the space naturally, replace electrical water heaters with solar-energy-based water heaters.

Zero-energy Building

Zero-energy test building in Tallinn, Estonia. Tallinn University of Technology.

A zero-energy building, also known as a zero net energy (ZNE) building, net-zero energy building (NZEB), or net zero building, is a building with zero net energy consumption, meaning the total amount of energy used by the building on an annual basis is roughly equal to the amount of renewable energy created on the site, or in other definitions by renewable energy sources elsewhere. These buildings consequently contribute less overall greenhouse gas to the atmosphere than similar non-ZNE buildings. They do at times consume non-renewable energy and produce greenhouse gases, but at other times reduce energy consumption and greenhouse gas production elsewhere by the same amount.

Most zero net energy buildings get half or more of their energy from the grid, and return the same amount at other times. Buildings that produce a surplus of energy over the year may be called "energy-plus buildings" and buildings that consume slightly more energy than they produce are called "near-zero energy buildings" or "ultra-low energy houses".

Traditional buildings consume 40% of the total fossil fuel energy in the US and European Union and are significant contributors of greenhouse gases. The zero net energy consumption principle is viewed as a means to reduce carbon emissions and reduce dependence on fossil fuels and although zero-energy buildings remain uncommon even in developed countries, they are gaining importance and popularity.

Most zero-energy buildings use the electrical grid for energy storage but some are independent of the grid. Energy is usually harvested on-site through energy producing technologies like solar and wind, while reducing the overall use of energy with highly efficient HVAC and lighting technologies. The zero-energy goal is becoming more practical as the costs of alternative energy technologies decrease and the costs of traditional fossil fuels increase.

The development of modern zero-energy buildings became possible not only through the progress made in new energy and construction technologies and techniques, but it has also been significantly improved by academic research, which collects precise energy performance data on traditional and experimental buildings and provides performance parameters for advanced computer models to predict the efficacy of engineering designs. Zero-energy buildings can be part of a smart grid. Some advantages of these buildings are as follows:

- Integration of renewable energy resources
- Integration of plug-in electric vehicles
- Implementation of zero-energy concepts

The net zero concept is applicable to a wide range of resources due to the many options for producing and conserving resources in buildings (e.g. energy, water, waste). Energy is the first resource to be targeted because it is highly managed, expected to continually become more efficient, and the ability to distribute and allocate it will improve disaster resiliency.

Definitions

Despite sharing the name "zero net energy", there are several definitions of what the term means in practice, with a particular difference in usage between North America and Europe.

Zero net site energy use

> In this type of ZNE, the amount of energy provided by on-site renewable energy sources is equal to the amount of energy used by the building. In the United States, "zero net energy building" generally refers to this type of building.

Zero net source energy use

> This ZNE generates the same amount of energy as is used, including the energy used to transport the energy to the building. This type accounts for losses during electricity transmission. These ZNEs must generate more electricity than zero net site energy buildings.

Net zero energy emissions

> Outside the United States and Canada, a ZEB is generally defined as one with zero net energy emissions, also known as a *zero carbon building* or *zero emissions building*. Under this definition the carbon emissions generated from on-site or off-site fossil fuel use are balanced by the amount of on-site renewable energy production. Other definitions include not only the carbon emissions gen-

erated by the building in use, but also those generated in the construction of the building and the embodied energy of the structure. Others debate whether the carbon emissions of commuting to and from the building should also be included in the calculation.Recent work in New Zealand has initiated an approach to include building user transport energy within zero energy building frameworks.

Net zero cost

In this type of building, the cost of purchasing energy is balanced by income from sales of electricity to the grid of electricity generated on-site. Such a status depends on how a utility credits net electricity generation and the utility rate structure the building uses.

Net off-site zero energy use

A building may be considered a ZEB if 100% of the energy it purchases comes from renewable energy sources, even if the energy is generated off the site.

Off-the-grid

Off-the-grid buildings are stand-alone ZEBs that are not connected to an off-site energy utility facility. They require distributed renewable energy generation and energy storage capability (for when the sun is not shining, wind is not blowing, etc.). An energy autarkic house is a building concept where the balance of the own energy consumption and production can be made on an hourly or even smaller basis. Energy autarkic houses can be taken off-the-grid.

Net zero-energy building

Based on scientific analysis within the joint research program "Towards Net Zero Energy Solar Buildings" a methodological framework was set up which allows different definitions, in accordance with country's political targets, specific (climate) conditions and respectively formulated requirements for indoor conditions: The overall conceptual understanding of a Net ZEB is an energy efficient, grid connected building enabled to generate energy from renewable sources to compensate its own energy demand.

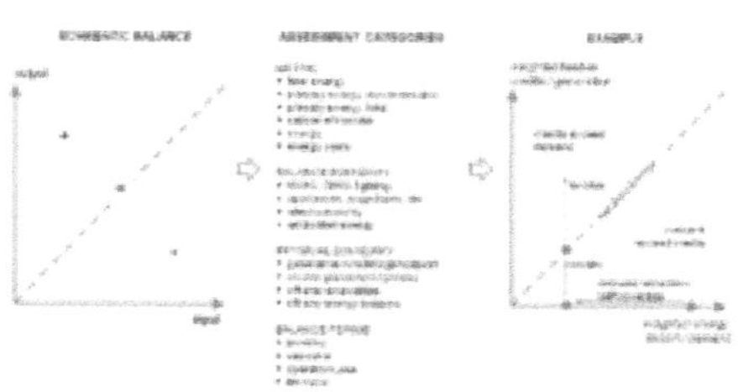

Figure 1: The Net ZEB balance concept: balance of weighted energy import respectively energy demand (x-axis) and energy export (feed-in credits) respectively (on-site) generation (y-axis)

> The wording "Net" emphasizes the energy exchange between the building and the energy infrastructure. By the building-grid interaction, the Net ZEBs becomes an active part of the renewable energy infrastructure. This connection to energy grids prevents seasonal energy storage and oversized on-site systems for energy generation from renewable sources like in energy autonomous buildings. The similarity of both concepts is a pathway of two actions: 1) reduce energy demand by means of energy efficiency measures and passive energy use; 2) generate energy from renewable sources. However, the Net ZEBs grid interaction and plans to widely increase their numbers evoke considerations on increased flexibility in the shift of energy loads and reduced peak demands.

Within this balance procedure several aspects and explicit choices have to be determined:

- The building system boundary is split into a physical boundary which determines which renewable resources are considered (e.g. in buildings footprint, on-site or even off-site) respectively how many buildings are included in the balance (single building, cluster of buildings) and a balance boundary which determines the included energy uses (e.g. heating, cooling, ventilation, hot water, lighting, appliances, IT, central services, electric vehicles, and embodied energy, etc.). It should be noticed that renewable energy supply options can be prioritized (e.g. by transportation or conversion effort, availability over the lifetime of the building or replication potential for future, etc.) and therefore create a hierarchy. It may be argued that resources within the building footprint or on-site should be given priority over off-site supply options.
- The weighting system converts the physical units of different energy carriers into a uniform metric (site/final energy, source/primary energy renewable parts included or not, energy cost, equivalent carbon emissions and even energy or environmental credits) and allows their comparison and compensation among each other in one single balance (e.g. exported PV electricity can compensate imported biomass). Politically influenced and therefore possibly asymmetrically or time dependent conversion/weighting factors can affect the relative value of energy carriers and can influence the required energy generation capacity.
- The balancing period is often assumed to be one year (suitable to cover all operation energy uses). A shorter period (monthly or seasonal) could also be considered as well as a balance over the entire life cycle (including embodied energy, which could also be annualized and counted in addition to operational energy uses).
- The energy balance can be done in two balance types: 1) Balance of delivered/imported and exported energy (monitoring phase as self-consumption of energy generated on-site can be included); 2) Balance between (weighted) energy

demand and (weighted) energy generation (for design phase as normally end users temporal consumption patterns -e.g. for lighting, appliances, etc.- are lacking). Alternatively a balance based on monthly net values in which only residuals per month are summed up to an annual balance is imaginable. This can be seen either as a load/generation balance or as a special case of import/export balance where a "virtual monthly self-consumption" is assumed.

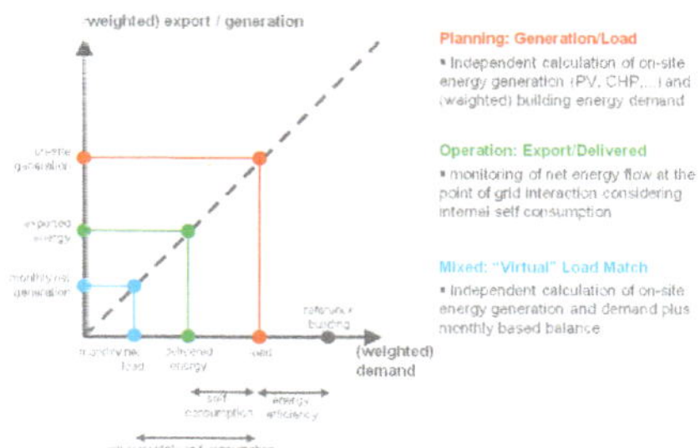

Figure 2: The Net ZEB balance concept: Graphical representation of the different types of balance: import/export balance between weighted exported and delivered energy, load/generation balance between weighted generation and load, and monthly net balance between weighted monthly net values of generation and load

- Beside the energy balance, Net ZEBs can be characterized by their ability to match the building's load by its energy generation (load matching) or to work beneficially with respect to the needs of the local grid infrastructure (grind interaction). Both can be expressed by suitable indicators which are intended as assessment tools only.

The information is based on the publications, and in which deeper information could be found.

Design and Construction

The most cost-effective steps toward a reduction in a building's energy consumption usually occur during the design process. To achieve efficient energy use, zero energy design departs significantly from conventional construction practice. Successful zero energy building designers typically combine time tested passive solar, or artificial conditioning, principles that work with the on-site assets. Sunlight and solar heat, prevailing breezes, and the cool of the earth below a building, can provide daylighting and stable indoor temperatures with minimum mechanical means. ZEBs are normally optimized to use passive solar heat gain and shading, combined with thermal mass to stabilize diurnal temperature variations throughout the day, and in most climates are superinsulated. All the technologies needed to create zero energy buildings are available off-the-shelf today.

Sophisticated 3-D building energy simulation tools are available to model how a building will perform with a range of design variables such as building orientation (relative to the daily and seasonal position of the sun), window and door type and placement, overhang depth, insulation type and values of the building elements, air tightness (weatherization), the efficiency of heating, cooling, lighting and other equipment, as well as local climate. These simulations help the designers predict how the building will perform before it is built, and enable them to model the economic and financial implications on building cost benefit analysis, or even more appropriate – life cycle assessment.

Zero-energy buildings are built with significant energy-saving features. The heating and cooling loads are lowered by using high-efficiency equipment, added insulation, high-efficiency windows, natural ventilation, and other techniques. These features vary depending on climate zones in which the construction occurs. Water heating loads can be lowered by using water conservation fixtures, heat recovery units on waste water, and by using solar water heating, and high-efficiency water heating equipment. In addition, daylighting with skylights or solartubes can provide 100% of daytime illumination within the home. Nighttime illumination is typically done with fluorescent and LED lighting that use 1/3 or less power than incandescent lights, without adding unwanted heat. And miscellaneous electric loads can be lessened by choosing efficient appliances and minimizing phantom loads or standby power. Other techniques to reach net zero (dependent on climate) are Earth sheltered building principles, superinsulation walls using straw-bale construction, Vitruvianbuilt pre-fabricated building panels and roof elements plus exterior landscaping for seasonal shading.

Zero-energy buildings are often designed to make dual use of energy including white goods; for example, using refrigerator exhaust to heat domestic water, ventilation air and shower drain heat exchangers, office machines and computer servers, and body heat to heat the building. These buildings make use of heat energy that conventional buildings may exhaust outside. They may use heat recovery ventilation, hot water heat recycling, combined heat and power, and absorption chiller units.

Energy Harvest

ZEBs harvest available energy to meet their electricity and heating or cooling needs. In the case of individual houses, various microgeneration technologies may be used to provide heat and electricity to the building, using solar cells or wind turbines for electricity, and biofuels or solar thermal collectors linked to a seasonal thermal energy storage (STES) for space heating. An STES can also be used for summer cooling by storing the cold of winter underground. To cope with fluctuations in demand, zero energy buildings are frequently connected to the electricity grid, export electricity to the grid when there is a surplus, and drawing electricity when not enough electricity is being produced. Other buildings may be fully autonomous.

Energy harvesting is most often more effective (in cost and resource utilization) when done on a local but combined scale, for example, a group of houses, cohousing, local district, village, etc. rather than an individual basis. An energy benefit of such localized energy harvesting is the virtual elimination of electrical transmission and electricity distribution losses. These losses amount to about 7.2%–7.4% of the energy transferred. Energy harvesting in commercial and industrial applications should benefit from the topography of each location. The production of goods under net zero fossil energy consumption requires locations of geothermal, microhydro, solar, and wind resources to sustain the concept.

Zero-energy neighborhoods, such as the BedZED development in the United Kingdom, and those that are spreading rapidly in California and China, may use distributed generation schemes. This may in some cases include district heating, community chilled water, shared wind turbines, etc. There are current plans to use ZEB technologies to build entire off-the-grid or net zero energy use cities.

The “Energy Harvest” Versus “Energy Conservation” Debate

One of the key areas of debate in zero energy building design is over the balance between energy conservation and the distributed point-of-use harvesting of renewable energy (solar energy, wind energy and thermal energy). Most zero energy homes use a combination of these strategies.

As a result of significant government subsidies for photovoltaic solar electric systems, wind turbines, etc., there are those who suggest that a ZEB is a conventional house with distributed renewable energy harvesting technologies. Entire additions of such homes have appeared in locations where photovoltaic (PV) subsidies are significant, but many so called “Zero Energy Homes” still have utility bills. This type of energy harvesting without added energy conservation may not be cost effective with the current price of electricity generated with photovoltaic equipment (depending on the local price of power company electricity), and may also requires greater embodied energy and greater resources so be thus the less ecological approach.

Since the 1980s, passive solar building design and passive house have demonstrated heating energy consumption reductions of 70% to 90% in many locations, without active energy harvesting. For new builds, and with expert design, this can be accomplished with little additional construction cost for materials over a conventional building. Very few industry experts have the skills or experience to fully capture benefits of the passive design. Such passive solar designs are much more cost-effective than adding expensive photovoltaic panels on the roof of a conventional inefficient building. A few kilowatt-hours of photovoltaic panels (costing 2 to 3 dollars per annual kWh production, U.S. dollar equivalent) may only reduce external energy requirements by 15% to 30%. A 100,000 BTU (110 MJ) high seasonal energy efficiency ratio 14 conventional air conditioner requires over 7 kW of photovoltaic electricity while it is operating, and that does not include enough for off-the-grid night-time operation. Passive cooling, and superior system engineering techniques, can reduce the air conditioning requirement by 70% to 90%. Photovoltaic-generated electricity becomes more cost-effective when the overall demand for electricity is lower.

Occupant Behavior

The energy used in a building can vary greatly depending on the behavior of its occupants. The acceptance of what is considered comfortable varies widely. Studies of identical homes in the United States have shown dramatic differences in energy use, with some homes using more than twice the energy of others. Occupant behavior can vary from differences in setting and programming thermostats, varying levels of illumination and hot water, and the amount of miscellaneous electric devices or plug loads used.

Utility Concerns

Utility companies are typically legally responsible for maintaining the electrical infrastructure that brings power to our cities, neighborhoods, and individual buildings. Utility companies typically own this infrastructure up to the property line of an individual parcel, and in some cases own electrical infrastructure on private land as well. Utilities have expressed concern that the use of Net Metering for ZNE projects threatens the Utilities base revenue, which in turn impacts their ability to maintain and service the portion of the electrical grid that they are responsible for. Utilities have expressed concern that states that maintain Net Metering laws may saddle non-ZNE homes with higher utility costs, as those homeowners would be responsible for paying for grid maintenance while ZNE home owners would theoretically pay nothing if they do achieve ZNE status. This creates potential equity issues, as currently, the burden would appear to fall on lower-income households. A possible solution to this issue is to create a minimum base charge for all homes connected to the utility grid, which would force ZNE home owners to pay for grid services independently of their electrical use.

Additional concerns exist that local distribution as well as larger transmission grids have not been designed to convey electricity in two directions, which may be necessary as higher levels of distributed energy generation come on line. Overcoming this barrier could require extensive upgrades to the electrical grid, however this is not believed to be a major problem until renewable generation reaches much higher levels of penetration than currently realized.

Development Efforts

Wide acceptance of zero-energy building technology may require more government incentives or building code regulations, the development of recognized standards, or significant increases in the cost of conventional energy.

The Google photovoltaic campus and the Microsoft 480-kilowatt photovoltaic campus relied on U.S. Federal, and especially California, subsidies and financial incentives. California is now providing US$3.2 billion in subsidies for residential-and-commercial near-zero-energy buildings, due to California's serious electricity shortage, frequent power outages, and air pollution problems. The details of other American states' renewable energy subsidies (up to US$5.00 per watt) can be found in the Database of State Incentives for Renewables and Efficiency. The Florida Solar Energy Center has a slide presentation on recent progress in this area.

The World Business Council for Sustainable Development has launched a major initiative to support the development of ZEB. Led by the CEO of United Technologies and the Chairman of Lafarge, the organization has both the support of large global companies and the expertise to mobilize the corporate world and governmental support to make ZEB a reality. Their first report, a survey of key players in real estate and construction, indicates that the costs of building green are overestimated by 300 percent. Survey respondents estimated that greenhouse gas emissions by buildings are 19 percent of the worldwide total, in contrast to the actual value of roughly 40 percent.

Influential zero-energy and Low-energy Buildings

Those who commissioned construction of passive houses and zero-energy homes (over the last three decades) were essential to iterative, incremental, cutting-edge, technology innovations. Much has been learned from many significant successes, and a few expensive failures.

The zero-energy building concept has been a progressive evolution from other low-energy building designs. Among these, the Canadian R-2000 and the German *passive house* standards have been internationally influential. Collaborative government demonstration projects, such as the superinsulated Saskatchewan House, and the International Energy Agency's *Task 13*, have also played their part.

Net Zero Energy Building Definition

The US National Renewable Energy Lab (NREL) published a groundbreaking report titled Net-Zero Energy Buildings: A Classification System Based on Renewable Energy Supply Options. This is the first report to lay out a full spectrum classification system for Net Zero/Renewable Energy buildings that includes the full spectrum of Clean Energy sources, both on site and off site. This classification system identifies the following 4 main categories of Net Zero Energy Buildings/Sites/Campuses:

- NZEB:A -- A footprint renewables Net Zero Energy Building
- NZEB:B -- A site renewables Net Zero Energy Building
- NZEB:C -- An imported renewables Net Zero Energy Building

- NZEB:D -- An off-site purchased renewables Net Zero Energy Building

Applying this U.S. Government Net Zero classification system means that every building "can" become Net Zero with the right combination of the key Net Zero Technologies - PV (solar), GHP (geothermal heating and cooling, thermal batteries), EE (energy efficiency), sometimes Wind, and Electric Batteries. A graphical exposé of the scale of impact of applying these NREL guidelines for Net Zero can be seen in the graphic at Net Zero Foundation titled "Net Zero Effect on U.S. Total Energy Use" showing a possible 39% U.S. total fossil fuel use reduction by changing U.S. Residential and Commercial buildings to Net Zero, 37% savings if we still use Nat. Gas for cooking at the same level.

Net Zero Carbon Conversion Example

Many well known universities have professed to want to completely convert their energy systems off of fossil fuels. The very idea that one could convert a whole campus off of fossil fuels has to date only been theoretical. Capitalizing on the continuing developments in both Photovoltaics and Geothermal heat pump technologies, and in the advancing Electric Battery field, complete conversion to a carbon free energy solution is now possible. An example of this is in the Net Zero Foundation's proposal at MIT to take that campus completely off fossil fuel use. This proposal shows the coming application of Net Zero Energy Buildings technologies at the District Energy scale.

Advantages and disadvantages

Advantages

- isolation for building owners from future energy price increases
- increased comfort due to more-uniform interior temperatures (this can be demonstrated with comparative isotherm maps)
- reduced requirement for energy austerity
- reduced total cost of ownership due to improved energy efficiency
- reduced total net monthly cost of living
- reduced risk of loss from grid blackouts
- improved reliability – photovoltaic systems have 25-year warranties and seldom fail during weather problems – the 1982 photovoltaic systems on the Walt Disney World EPCOT Energy Pavilion are still working fine today, after going through three recent hurricanes
- extra cost is minimized for new construction compared to an afterthought retrofit
- higher resale value as potential owners demand more ZEBs than available supply

- the value of a ZEB building relative to similar conventional building should increase every time energy costs increase
- future legislative restrictions, and carbon emission taxes/penalties may force expensive retrofits to inefficient buildings

Disadvantages

- initial costs can be higher – effort required to understand, apply, and qualify for ZEB subsidies, if they exist.
- very few designers or builders have the necessary skills or experience to build ZEBs
- possible declines in future utility company renewable energy costs may lessen the value of capital invested in energy efficiency
- new photovoltaic solar cells equipment technology price has been falling at roughly 17% per year – It will lessen the value of capital invested in a solar electric generating system – Current subsidies will be phased out as photovoltaic mass production lowers future price
- challenge to recover higher initial costs on resale of building, but new energy rating systems are being introduced gradually.
- while the individual house may use an average of net zero energy over a year, it may demand energy at the time when peak demand for the grid occurs. In such a case, the capacity of the grid must still provide electricity to all loads. Therefore, a ZEB may not reduce the required power plant capacity.
- without an optimised thermal envelope the embodied energy, heating and cooling energy and resource usage is higher than needed. ZEB by definition do not mandate a minimum heating and cooling performance level thus allowing oversized renewable energy systems to fill the energy gap.
- solar energy capture using the house envelope only works in locations unobstructed from the sun. The solar energy capture cannot be optimized in north (for northern hemisphere, or south for southern Hemisphere) facing shade, or wooded surroundings.

Zero Energy Building Versus Green building

The goal of green building and sustainable architecture is to use resources more efficiently and reduce a building's negative impact on the environment. Zero energy buildings achieve one key green-building goal of completely or very significantly reducing energy use and greenhouse gas emissions for the life of the building. Zero energy buildings may or may not be considered "green" in all areas, such as reducing waste, using recycled building materials, etc. However, zero energy, or net-zero buildings do tend to have a much lower ecological impact over the life of the building compared with other "green" buildings that require imported energy and/or fossil fuel to be habitable and meet the needs of occupants.

Because of the design challenges and sensitivity to a site that are required to efficiently meet the energy needs of a building and occupants with renewable energy (solar, wind, geothermal, etc.), designers must apply holistic design principles, and take advantage of the free naturally occurring assets available, such as passive solar orientation, natural ventilation, daylighting, thermal mass, and night time cooling.

Certification

Many green building certification programs do not require a building to have net zero energy use, only to reduce energy use a few percentage points below the minimum required by law. Green Globes involves check lists that are measurement tools, not design tools. Inexperienced designers or architects may cherry-pick points to meet a target certification level, even though those points may not be the best design choices for a specific building or climate. In November, 2011, the International Living Future Institute developed the Net Zero Energy Building Certification. Designed as part of the Living Building Challenge, Net Zero Energy Building Certification is simple, cost effective and critical for integrity and transparency.

Worldwide

International Initiatives

Between 2008 and 2013, researchers from Australia, Austria, Belgium, Canada, Denmark, Finland, France, Germany, Italy, Republic of Korea, New Zealand, Norway, Portugal, Singapore, Spain, Sweden, Switzerland, United Kingdom and USA were working together in the joint research program "Towards Net Zero Energy Solar Buildings" under the umbrella of International Energy Agency (IEA) Solar Heating and Cooling Program (SHC) Task 40 / Energy in Buildings and Communities (EBC, formerly ECBCS) Annex 52 in order to bring the Net ZEB concept to market viability. The joint international research and demonstration activities are divided in subtasks. The objective is to develop a common understanding, a harmonized international applicable definition framework, design process tools (Subtask B), advanced building design and technology solutions and industry guidelines for Net ZEBs (Subtask C). The scope encompasses new and existing residential and non-residential buildings located within the climatic zones of the participating countries.

Australia

In Australia, researchers have recently developed a new approach to the construction of visually-clear solar energy harvesting windows suitable for industrialization and applications in net-zero energy buildings. Industrial production of several prototype batches of solar windows has started in 2016.

Belgium

In Belgium there is a project with the ambition to make the Belgian city Leuven climate-neutral in 2030.

Japan

After April 2011 Fukushima earthquake follow up with Fukushima Daiichi nuclear disaster, Japan experienced severe power crisis that led to the awareness of importance of energy conservation. In 2012 Ministry of Economy, Trade and Industry, Ministry of Land, Infrastructure, Transport and Tourism and Ministry of the Environment (Japan) summarized the road map for Low-carbon Society which contains the goal of ZEH and ZEB to be standard of new construction in 2020.

Canada

- On May 3, 2013, Prime Minister Harper announced funding for ecoENERGY Innovation Initiative projects including a project being led by Owens Corning entitled Integrating Renewables and Conservation Measures in a Net-Zero Energy Low-Rise Residential Subdivision. This demonstration project is aimed at addressing challenges specific to production housing when building to net zero energy performance levels. The buildABILITY Corporation project management team will be working to assess and resolve challenges in relation to site planning, construction, equipment, grid connections, cost, trade capability, warranty, reliability, sales, marketing, and homebuyer information/education. Five home builders across four provinces will build at least 25 Net Zero Energy (NZE) homes by March 2016 as part of this project. The five selected builders participating in this initiative are: Mattamy Homes Limited (Calgary, Alberta); Construction Voyer (Laval, Quebec); Minto Communities (Ottawa, Ontario); Provident Development Inc. (Halifax, Nova Scotia); and Reid's Heritage Homes (Guelph, Ontario).
- In Canada the Net-Zero Energy Home Coalition is an industry association promoting net-zero energy home construction and the adoption of a near net-zero energy home (nNZEH), NZEH Ready and NZEH standard.
- The Canada Mortgage and Housing Corporation is sponsoring the EQuilibrium Sustainable Housing Competition that will see the completion of fifteen zero-energy and near-zero-energy demonstration projects across the country starting in 2008.

- The EcoTerra House in Eastman, Quebec is Canada's first nearly net-zero energy housing built through the CMHC EQuilibrium Sustainable Housing Competition. The house was designed by Assoc. Prof. Dr. Masa Noguchi of the University of Melbourne for Alouette Homes and engineered by Prof. Dr. Andreas K. Athienitis of Concordia University.
- The EcoPlusHome in Bathurst, New Brunswick. The Eco Plus Home is a prefabricated test house built by Maple Leaf Homes and with technology from Bosch Thermotechnology.
- The first net-zero passive house in Northshore, Vancouver, BC, is designed by Dr. Homayoun Arbabian. The design and construction of this SuperEcoHouse is undertaken by Vancouver Green Homes LTD.

China

- One example of the new generation of zero energy office buildings is the 71-story Pearl River Tower, which opened in 2009, as the China National Tobacco Corporation headquarters. It uses both modest energy efficiency, and a big distributed renewable energy generation from both solar and wind. Designed by Skidmore Owings Merrill LLP in Guangzhou, China, the tower is receiving economic support from government subsidies that are now funding many significant conventional fossil-fuel (and nuclear energy) energy reduction efforts.
- Dongtan Eco-City near Shanghai

Denmark

Strategic Research Centre on Zero Energy Buildings was in 2009 established at Aalborg University by a grant from the Danish Council for Strategic Research (DSF), the Programme Commission for Sustainable Energy and Environment, and in cooperation with the Technical University of Denmark, Danish Technological Institute, Danfoss A/S, Velux A/S, Saint Gobain Isover A/S, and The Danish Construction Association, the section of aluminium facades. The purpose of the centre is through development of integrated, intelligent technologies for the buildings, which ensure considerable energy conservations and optimal application of renewable energy, to develop zero energy building concepts. In cooperation with the industry, the centre will create the necessary basis for a long-term sustainable development in the building sector.

Germany

- Technische Universität Darmstadt won first place in the international zero energy design 2007 Solar Decathlon competition, with a passivhaus design (Passive house) + renewables, scoring highest in the Architecture, Lighting, and Engineering contests

- Fraunhofer Institute for Solar Energy Systems ISE, Freiburg im Breisgau
- Net zero energy- , energy-plus or climate-neutral buildings in the next generation of electricity grids

India

India's first net zero building is Indira Paryavaran Bhawan, located in New Delhi. Features include passive solar building design and other green technologies.

Iran

In 2011, *Payesh* Energy House (PEH) or *Khaneh Payesh Niroo* by a collaboration of Fajr-e-Toseah Consultant Engineering Company and Vancouver Green Homes Ltd] under management of Payesh Energy Group (EPG) launched the first Net-Zero passive house in Iran. This concept makes the design and construction of PEH a sample model and standardized process for mass production by MAPSA.

Also an example of the new generation of zero energy office buildings is the 24-story OIIC Office Tower, which is started in 2011, as the OIIC Company headquarters. It uses both modest energy efficiency, and a big distributed renewable energy generation from both solar and wind. It is managed by Rahgostar Naft Company in Tehran, Iran. The tower is receiving economic support from government subsidies that are now funding many significant fossil-fuel-free efforts.

Ireland

In 2005, *Scandinavian Homes* launched the world's first standardised passive house in Ireland, this concept makes the design and construction of passive house a standardised process. Conventional low energy construction techniques have been refined and modelled on the PHPP (Passive House Design Package) to create the standardised passive house. Building offsite allows high precision techniques to be utilised and reduces the possibility of errors in construction.In 2009 the same company started a project to use 23,000 liters of water in a seasonal storage tank, heated up by evacuated solar tubes throughout the year, with the aim to provide the house with enough heat throughout the winter months thus eliminating the need for any electrical heat to keep the house comfortably warm. The system is monitored and documented by a research team from The University of Ulster and the results will be included in part of a PhD thesis.

In 2012 Cork institute of Technology started renovation work on its 1974 building stock to develop a net zero energy building retrofit. The exemplar project will become Ireland's first zero energy testbed offering a post occupancy evaluation of actual building performance against design benchmarks.

Malaysia

In October 2007, the Malaysia Energy Centre (PTM) successfully completed the development and construction of the PTM Zero Energy Office (ZEO) Building. The building has been designed to be a super-energy-efficient building using only 286 kWh/day. The renewable energy – photovoltaic combination is expected to result in a net zero energy requirement from the grid. The building is currently undergoing a fine tuning process by the local energy management team. Findings are expected to be published in a year.

Netherlands

In September 2006, the Dutch headquarters of the World Wildlife Fund (WWF) in Zeist was opened. This earth-friendly building gives back more energy than it uses. All materials in the building were tested against strict requirements laid down by the WWF and the architect.

Norway

In February 2009, the Research Council of Norway assigned The Faculty of Architecture and Fine Art at the Norwegian University of Science and Technology to host the Research Centre on Zero Emission Buildings (ZEB), which is one of eight new national Centres for Environment-friendly Energy Research (FME). The main objective of the FME-centres is to contribute to the development of good technologies for environmentally friendly energy and to raise the level of Norwegian expertise in this area. In addition, they should help to generate new industrial activity and new jobs. Over the next eight years, the FME-Centre ZEB will develop competitive products and solutions for existing and new buildings that will lead to market penetration of zero emission buildings related to their production, operation and demolition.

Singapore

Singapore's first zero-energy building was launched at the inaugural Singapore Green Building Week.

Switzerland

The Swiss MINERGIE-A-Eco label certifies zero energy buildings. The first building with this label, a single-family home, was completed in Mühleberg in 2011.

United Arab Emirates

- Masdar City in Abu Dhabi

United Kingdom

In December 2006, the government announced that by 2016 all new homes in England will be zero energy buildings. To encourage this, an exemption from Stamp Duty Land Tax is planned. In Wales the plan is for the standard to be met earlier in 2011, although it is looking more likely that the actual implementation date will be 2012. However, as a result of a unilateral change of policy published at the time of the March 2011 budget, a more limited policy is now planned which, it is estimated, will only mitigate two thirds of the emissions of a new home.

- BedZED development
- Hockerton Housing Project

United States

Figure 3: Net Zero Court zero emissions office building prototype in St. Louis, Missouri

In the US, ZEB research is currently being supported by the US Department of Energy (DOE) Building America Program, including industry-based consortia and researcher organizations at the National Renewable Energy Laboratory (NREL), the Florida Solar Energy Center (FSEC), Lawrence Berkeley National Laboratory (LBNL), and Oak Ridge National Laboratory (ORNL). From fiscal year 2008 to 2012, DOE plans to award $40 million to four Building America teams, the Building Science Corporation; IBACOS; the Consortium of Advanced Residential Buildings; and the Building Industry Research Alliance, as well as a consortium of academic and building industry leaders. The funds will be used to develop net-zero-energy homes that consume at 50% to 70% less energy than conventional homes.

DOE is also awarding $4.1 million to two regional building technology application centers that will accelerate the adoption of new and developing energy-efficient technologies. The two centers, located at the University of Central Florida and Washington State University, will serve 17 states, providing information and training on commercially available energy-efficient technologies.

The U.S. Energy Independence and Security Act of 2007 created 2008 through 2012 funding for a new solar air conditioning research and development program, which should soon demonstrate multiple new technology innovations and mass production economies of scale.

The 2008 Solar America Initiative funded research and development into future development of cost-effective Zero Energy Homes in the amount of $148 million in 2008.

The Solar Energy Tax Credits have been extended until the end of 2016. Solar power in the United States

By Executive Order 13514, U.S. President Barack Obama mandated that by 2015, 15% of existing Federal buildings conform to new energy efficiency standards and 100% of all new Federal buildings be Zero-Net-Energy by 2030.

Energy Free Home Challenge

In 2007, the philanthropic Siebel Foundation created the Energy Free Home Foundation. The goal was to offer $20 million in global incentive prizes to design and build a 2,000 square foot (186 square meter) three-bedroom, two bathroom home with (1) net-zero annual utility bills that also has (2) high market appeal, and (3) costs no more than a conventional home to construct.

The plan included funding to build the top ten entries at $250,000 each, a $10 million first prize, and then a total of 100 such homes to be built and sold to the public.

Beginning in 2009, Thomas Siebel made many presentations about his Energy Free Home Challenge. The Siebel Foundation Report stated that the Energy Free Home Challenge was "Launching in late 2009".

The Lawrence Berkeley National Laboratory at the University of California, Berkeley participated in writing the "Feasibility of Achieving Zero-Net-Energy, Zero-Net-Cost Homes" for the $20-million Energy Free Home Challenge.

If implemented, the Energy Free Home Challenge would have provided increased incentives for improved technology and consumer education about zero energy buildings coming in at the same cost as conventional housing.

U.S. Department of Energy Solar Decathlon

The U.S. Department of Energy Solar Decathlon is an international competition that challenges collegiate teams to design, build, and operate the most attractive, effective, and energy-efficient solar-powered house. Achieving Zero Net Energy balance is a major focus of the competition.

List

Arizona

- Zero Energy House developed by the NAHB Research Center and John Wesley Miller Companies, Tucson.

California

- The State of California has proposed that all new low- and mid-rise residential buildings, and all new commercial buildings, be designed and constructed to ZNE standards beginning in 2020 and 2030, respectively. The requirements, if implemented, will be promulgated via the California Building Code, which is updated on a three-year cycle and which currently mandates some of the highest energy efficiency standards in the United States. California is anticipated to further increase efficiency requirements by 2020, thus avoiding the trends discussed above of building standard housing and achieving ZNE by adding large amounts of renewables. The California Energy Commission is required to perform a cost-benefit analysis to prove that new regulations create a net benefit for residents of the state. There has yet to be publicly released analysis of the impact that ZNE standards may have to construction, real estate, and energy prices in the state. Debate exists as to if the state's stringent efficiency requirements are directly responsible for the apparent "flatlining" of residential electricity use in the state since 1975. This debate is relevant to ZNE codes, as it remains to be seen via models or in practice what overall effect the proliferation of ZNE buildings will have on overall electricity use in the state, and at what cost.

- West Village, located on the University of California campus in Davis, California, was the largest ZNE-planned community in North America at the time of its opening in 2014. The development contains student housing for approximately 1,980 UC Davis students as well as leasable office space and community amenities including a community center, pool, gym, restaurant and convenience store. Office spaces in the development are currently leased by energy and transportation-related University programs. The project was a public-private partnership between the university and West Village Community Partnership LLC, led by Carmel Partners of San Francisco, a private developer, who entered into a 60-year ground lease with the university and was responsible for the design, construction, and implementation of the $300 million project, which is intended to be market-rate housing for Davis. This is unique as the developer designed the project to achieve ZNE at no added cost to themselves or to the residents. Designed and modeled to achieve ZNE, the project uses a mixture of passive elements (roof overhangs, well-insulated walls, radiant heat barriers, ducts in insulated spaces, etc.) as well as active approaches (occupancy sensors on lights, high-efficiency appliances and lighting, etc.). Designed to out-perform California's 2008 Title 24 energy codes by 50%, the project produced 87% of the energy it consumed during its first year in operation. The shortcoming in ZNE status is attributed to several factors, including improperly functioning heat pump water heaters, which have since been fixed. Occupant behavior is significantly different than anticipated, with the all-student population using more energy on a per-capita basis than typical inhabitants of single-family homes in the area. One of the primary factors driving increased energy use appears to be the increased miscellaneous electrical loads (MEL, or plug loads) in the form of mini-refrigerators, lights, computers, gaming consoles, televisions, and other electronic equipment. The university continues to work with the developer to identify strategies for achieving ZNE status. These approaches include incentivizing occupant behavior and increasing the site's renewable energy capacity, which is a 4 MW photovoltaic array per the original design. The West Village site is also home to the Honda Smart Home US, a beyond-ZNE single-family home that incorporates cutting-edge technologies in energy management, lighting, construction, and water efficiency.

- The IDeAs Z2 Design Facility is a net zero energy, zero carbon retrofit project occupied since 2007. It uses less than one fourth the energy of a typical U.S. office by applying strategies such as daylighting, radiant heating/cooling with a ground-source heat pump and high energy performance lighting and computing. The remaining energy demand is met with renewable energy from its building-integrated photovoltaic array. In 2009, building owner and occupant Integrated Design Associates (IDeAs) recorded actual measured energy use intensity of 21.17 kbtu/sf-year, with 21.72 kbtu/sf-year produced, for a net of −0.55 kbtu/sf-yr. The building is also carbon neutral, with no gas connection, and with carbon offsets purchased to cover the embodied carbon of the building materials used in the renovation.

- The Zero Net Energy Center, scheduled to open in 2013 in San Leandro, is to be a 46,000-square-foot electrician training facility created by the International Brotherhood of Electrical Workers Local 595 and the Northern California chapter of the National Electrical Contractors Association. Training will include energy-efficient construction methods.

- The Green Idea House is a net zero energy, zero-carbon retrofit in Hermosa Beach.

- George LeyVa Middle School Administrative Offices, occupied since fall 2011, is a net zero energy, net zero carbon emissions building of just over 9,000 square feet. With daylighting, variable refrigerant flow HVAC, and displacement ventilation, it is designed to use half of the energy of a conventional California school building, and, through a building-integrated solar array, provides 108% of the energy needed to offset its annual electricity use. The excess helps power the remainder of the middle school campus. It is the first publicly funded NZE K–12 building in California.

Colorado

- The Moore House achieves net-zero energy usage with passive solar design, 'tuned' heat reflective windows, super-insulated and air-tight construction, natural daylighting, solar thermal panels for hot water and space heating, a photovoltaic (PV) system that generates more carbon-free electricity than the house requires, and an energy-recovery ventilator (ERV) for fresh air. The green building strategies used by Thomas Doerr of Doerr Architecture and Ecofutures Building on the Moore House earned it a verified home energy rating system (HERS) score of −3.
- The NREL Research Support Facility in Golden is an award-winning class A office building. Its energy efficiency features include: Thermal storage concrete structure, transpired solar collectors, 70 miles of radiant piping, high-efficiency office equipment, and an energy-efficient data center that reduces the data center's energy use by 50% over traditional approaches.
- Wayne Aspinall Federal Building in Grand Junction, originally constructed in 1918, became the first Net Zero Energy building listed on the National Register of Historic Places. On-site renewable energy generation is intended to produce 100% of the building's energy throughout the year using the following energy efficiency features: Variable refrigerant flow for the HVAC, a geo-exchange system, advanced metering and building controls, high-efficient lighting systems, thermally enhanced building envelope, interior window system (to maintain historic windows), and advanced power strips (APS) with individual occupancy sensors.

Florida

- The 1999 side-by-side Florida Solar Energy Center Lakeland demonstration project was called the "Zero Energy Home." It was a first-generation university effort that significantly influenced the creation of the U.S. Department of Energy, Energy Efficiency and Renewable Energy, Zero Energy Home program.

Illinois

- The Walgreens store located on 741 Chicago Ave, Evanston, is the first of the company's stores to be built and or converted to a net zero energy building. It is the first net zero energy retail stores to be built and will pave the way to renovating and building net zero energy retail stores in the near future. The Walgreens store includes the following energy efficiency features: Geo-exchange system, energy-efficient building materials, LED lighting and daylight harvesting, and carbon dioxide refrigerant.

- The Electrical and Computer Engineering building at the University of Illinois at Urbana-Champaign, which was built in 2014, is a net zero building.

Iowa

- The MUM Sustainable Living Center was designed to surpass LEED Platinum qualification. The Maharishi University of Management (MUM) in Fairfield, Iowa, founded by Maharishi Mahesh Yogi (best known for having brought Transcendental Meditation to the West) incorporates principles of Bau Biology (a German system that focuses on creating a healthy indoor environment), as well as Maharishi Vedic Architecture (an Indian system of architecture focused on the precise orientation, proportions and placement of rooms). The building is one of the few in the country to qualify as net zero, and one of even fewer that can claim the banner of grid positive via its solar power system. A rainwater catchment system and on-site natural waste-water treatment likewise take the building off (sewer) grid with respect to water and waste treatment. Additional green features include natural daylighting in every room, natural and breathable earth block walls (made by the program's students), purified rainwater for both potable and non-potable functions; and an on-site water purification and recycling system consisting of plants, algae, and bacteria.

Kentucky

- Richardsville Elementary School, part of the Warren County Public School District in south central Kentucky, is the first Net Zero energy school in the United States. To reach Net Zero, innovative energy reduction strategies were used by CMTA Consulting Engineers and Sherman Carter Barnhart Architects including dedicated outdoor air systems (DOAS) with dynamic reset, new IT systems, alternative methods to prepare lunches, and the use of solar photovoltaics. The project has an efficient thermal envelope constructed with insulated concrete form (ICF) walls, geothermal water source heat pumps, low-flow fixtures, and features daylighting extensively throughout. It is also the first truly wireless school in Kentucky.

- Locust Trace AgriScience Center, an agricultural-based vocational school serving Fayette County Public Schools and surrounding districts, features a Net Zero Academic Building engineered by CMTA Consulting Engineers and designed by Tate Hill Jacobs Architects. The facility, located in Lexington, Kentucky, also has a greenhouse, riding arena with stalls, and a barn. To reach Net Zero in the Academic Building the project utilizes an air-tight envelope, expanded indoor temperature setpoints in specified areas to more closely model real-world conditions, a solar thermal system, and geothermal water source heat pumps. The school has further reduced its site impact by minimizing municipal water use through the use of a dual system consisting of a standard leach field system and a constructed wetlands system and using pervious surfaces to collect, drain, and use rainwater for crop irrigation and animal watering.

Massachusetts

- The government of Cambridge has enacted a plan for "net zero" carbon emissions from all buildings in the city by 2040.
- John W. Olver Transit Center, designed by Charles Rose Architects Inc, is an intermodal transit hub in Greenfield, Massachusetts. Built with American Recovery and Reinvestment Act funds, the facility was constructed with solar panels, geothermal wells, copper heat screens and other energy efficient technologies.

Michigan

- The Mission Zero House is the 110-year-old Ann Arbor home of Greenovation. TV host and Environment Report contributor Matthew Grocoff. As of 2011, the home is the oldest home in America to achieve net-zero energy. The owners are chronicling their project on Greenovation.TV and the Environment Report on public radio.
- The Vineyard Project is a Zero Energy Home (ZEH) thanks to the Passive Solar Design, 3.3 Kws of Photovoltaics,Solar Hot Water and Geothermal Heating and Cooling. The home is pre-wired for a future wind turbine and only uses 600 kWh of energy per month while a minimum of 20 kWh of electricity per day with many days net-metering backwards. The project also used ICF insulation throughout the entire house and is certified as Platinum under the LEED for Homes certification. This Project was awarded Green Builder Magazine Home of the Year 2009
- The Lenawee Center for a Sustainable Future, a new campus for Lenawee Intermediate School District, serves as a living laboratory for the future of agriculture. It is the first Net Zero education building in Michigan, engineered by CMTA Consulting Engineers and designed by The Collaborative, Inc. The project includes solar arrays on the ground as well as the roof, a geothermal heating and cooling system, solar tubes, permeable pavement and sidewalks, a sedum green roof, and an overhang design to regulate building temperature.

Missouri

- In 2010, architectural firm HOK worked with energy and daylighting consultant The Weidt Group to design a 170,735-square-foot (15,861.8 m^2) net zero carbon emissions Class A office building prototype in St. Louis, Missouri. The team chronicled its process and results on Netzerocourt.com.

New Jersey

- The 31 Tannery Project, located in Branchburg, New Jersey, serves as the corporate headquarters for Ferreira Construction, the Ferreira Group, and Noveda Technologies. The 42,000-square-foot (3,900 m^2) office and shop building was constructed in 2006 and is the first building in the state of New Jersey to meet New Jersey's Executive Order 54. The building is also the first Net Zero Electric Commercial Building in the United States.

New York

- Green Acres, the first true zero-net energy development in America, is located in New Paltz, about 80 miles (130 km) north of New York City. Greenhill Contracting began construction on this development of 25 single family homes in summer 2008, with designs by BOLDER Architecture. After a full year of occupancy, from March 2009 to March 2010, the solar panels of the first occupied home in Green Acres generated 1490 kWh more energy than the home consumed. The second occupied home has also achieved zero-net energy use. As of June 2011, 5 houses have been completed, purchased and occupied, 2 are under construction, and several more are being planned. The homes are built of insulated concrete forms with spray foam insulated rafters and triple pane casement windows, heated and cooled by a geothermal system, to create extremely energy-efficient and long-lasting buildings. The heat recovery ventilator provides constant fresh air and, with low or no VOC (volatile organic compound) materials, these homes are very healthy to live in. To the best of our knowledge, Green Acres is the first development of multiple buildings, residential or commercial, that achieves true zero-net energy use in the United States, and the first zero-net energy development of single family homes in the world.

- Greenhill Contracting has built two luxury zero-net energy homes in Esopus, completed in 2008. One house was the first Energy Star rated zero-net energy home in the Northeast and the first registered zero-net energy home on the US Department of Energy's Builder's Challenge website. These homes were the template for Green Acres and the other zero-net energy homes that Greenhill Contracting has built, in terms of methods and materials.

- The headquarters of Hudson Solar, a dba of Hudson Valley Clean Energy, Inc., located in Rhinebeck and completed in 2007, was determined by NESEA (the Northeast Sustainable Energy Association) to have become the first proven zero-net energy commercial building in New York State and the ten northeast United States (October 2008). The building consumes less energy than it generates, using a solar electric system to generate power from the sun, geothermal heating and cooling, and solar thermal collectors to heat all its hot water.

Oklahoma

- The first 5,000-square-foot (460 m²) zero-energy design home was built in 1979 with support from President Carter's new United States Department of Energy. It relied heavily on passive solar building design for space heat, water heat and space cooling. It heated and cooled itself effectively in a climate where the summer peak temperature was 110 degrees Fahrenheit, and the winter low temperature was −10 F. It did not use active solar systems. It is a double envelope house that uses a gravity-fed natural convection air flow design to circulate passive solar heat from 1,000 square feet (93 m²) of south-facing glass on its greenhouse through a thermal buffer zone in the winter. A swimming pool in the greenhouse provided thermal mass for winter heat storage. In the summer, air from two 24-inch (610 mm) 100-foot-long (30 m) underground earth tubes is used to cool the thermal buffer zone and exhaust heat through 7200 cfm of outer-envelope roof vents.

Oregon

- Net Zero Energy Building Certification launched in 2011, with an international following. The first project, Painters Hall, is Pringle Creek's Community Center, café, office, art gallery, and event venue. Originally built in the 1930s, Painters Hall was renovated to LEED Platinum Net Zero energy building standards in 2010, demonstrating the potential of converting existing building stock into high-performance, sustainable building sites. Painters Hall features simple low--cost solutions for energy reduction, such as natural daylighting and passive cooling lighting, that save money and increase comfort. A district ground-source geothermal loop serves the building's GSHP for highly efficient heating and air conditioning. Excess generation from the 20.2 kW rooftop solar array offsets pumping for the neighborhoods geo loop system. Open to the public, Painters Hall is a hub for gatherings of friends, neighbors, and visitors at the heart of a neighborhood designed around nature and community.

Pennsylvania

- The Phipps Center for Sustainable Landscapes in Pittsburgh was designed to be one of the greenest buildings in the world. It achieved Net Zero Energy Building Certification from the Living Building Challenge in February 2014 and is pursuing full certification. The Phipps Center uses energy conservation technologies such as solar hot water collectors, carbon dioxide sensors, and daylighting, as well as renewable energy technologies to allow it to achieve Net Zero Energy status.

Rhode Island

- In Newport, the Paul W. Crowley East Bay MET School is the first Net Zero project to be constructed in Rhode Island. It is a 17,000 sq ft building, housing eight large classrooms, seven bathrooms and a kitchen. It will have PV panels to supply all necessary electricity for the building and a geothermal well which will be the source of heat.

Figure 4: Zero-Energy Lab construction on UNT campus in Denton, Texas

Texas

- The University of North Texas (UNT) was constructing a Zero Energy Research Laboratory on its 300-acre research campus, Discovery Park, in Denton, Texas. The project is funded at over $1,150,000 and will primarily benefit students in mechanical and energy engineering (UNT became the first university to offer degrees in mechanical and energy engineering in 2006). This 1,200-square-foot structure is now competed and held ribbon cutting ceremony for the University of North Texas' Zero Energy Laboratory on April 20, 2012.

Vermont

- The Putney School's net zero Field House was opened October 10, 2009. In use for over a year, as of December 2010, the Field House used 48,374 kWh and produced a total of 51,371 kWh during the first 12 months of operation, thus performing at slightly better than net-zero. Also in December, the building won an AIA-Vermont Honor Award.
- The Charlotte Vermont House designed by Pill-Maharam Architects is a verified net zero energy house completed in 2007. The project won the Northeast Sustainable Energy Association's Net Zero Energy award in 2009.

Passive Cooling

Passive cooling is a building design approach that focuses on heat gain control and heat dissipation in a building in order to improve the indoor thermal comfort with low or nil energy consumption. This approach works either by preventing heat from entering the interior (heat gain prevention) or by removing heat from the building (natural cooling). Natural cooling utilizes on-site energy, available from the natural environment, combined with the architectural design of building components (e.g. building envelope), rather than mechanical systems to dissipate heat. Therefore, natural cooling depends not only on the architectural design of the building but on how the site's natural resources are used as heat sinks (i.e. everything that absorbs or dissipates heat). Examples of on-site heat sinks are the upper atmosphere (night sky), the outdoor air (wind), and the earth/soil.

Overview

Passive cooling covers all natural processes and techniques of heat dissipation and modulation without the use of energy. Some authors consider that minor and simple mechanical systems (e.g. pumps and economizers) can be integrated in passive cooling techniques, as long they are used to enhance the effectiveness of the natural cooling process. Such applications are also called 'hybrid cooling systems'. The techniques for passive cooling can be grouped in two main categories:

- Preventative techniques that aims to provide protection and/or prevention of external and internal heat gains.
- Modulation and heat dissipation techniques allow the building to store and dissipate heat gain through the transfer of heat from heat sinks to the climate. This technique can be the result of thermal mass or natural cooling.

Preventative Techniques

Protection from or prevention of heat gains encompasses all the design techniques that minimizes the impact of solar heat gains through the building's envelope and of internal heat gains that is generated inside the building due occupancy and equipment. It includes the following design techniques:

- Microclimate and site design - By taking into account the local climate and the site context, specific cooling strategies can be selected to apply which are the most appropriate for preventing overheating through the envelope of the building. The microclimate can play a huge role in determining the most favorable building location by analyzing the combined availability of sun and wind. The bioclimatic chart, the solar diagram and the wind rose are relevant analysis tools in the application of this technique.
- Solar control - A properly designed shading system can effectively contribute to minimizing the solar heat gains. Shading both transparent and opaque surfaces of the building envelope will minimize the amount of solar radiation that induces overheating in both indoor spaces and building's structure. By shading the building structure, the heat gain captured through the windows and envelope will be reduced.

- Building form and layout - Building orientation and an optimized distribution of interior spaces can prevent overheating. Rooms can be zoned within the buildings in order to reject sources of internal heat gain and/or allocating heat gains where they can be useful, considering the different activities of the building. For example, creating a flat, horizontal plan will increase the effectiveness of cross-ventilation across the plan. Locating the zones vertically can take advantage of temperature stratification. Typically, building zones in the upper levels are warmer than the lower zones due to stratification. Vertical zoning of spaces and activities uses this temperature stratification to accommodate zone uses according to their temperature requirements. Form factor (i.e. the ratio between volume and surface) also plays a major role in the building's energy and thermal profile. This ratio can be used to shape the building form to the specific local climate. For example, more compact forms tend to preserve more heat than less compact forms because the ratio of the internal loads to envelope area is significant.

- Thermal insulation - Insulation in the building's envelope will decrease the amount of heat transferred by radiation through the facades. This principle applies both to the opaque (walls and roof) and transparent surfaces (windows) of the envelope. Since roofs could be a larger contributor to the interior heat load, especially in lighter constructions (e.g. building and workshops with roof made out of metal structures), providing thermal insulation can effectively decrease heat transfer from the roof.

- Behavioral and occupancy patterns - Some building management policies such as limiting the number of people in a given area of the building can also contribute effectively to the minimization of heat gains inside a building. Building occupants can also contribute to indoor overheating prevention by: shutting off the lights and equipment of unoccupied spaces, operating shading when necessary to reduce solar heat gains through windows, or dress lighter in order to adapt better to the indoor environment by increasing their thermal comfort tolerance.

- Internal gain control - More energy-efficient lighting and electronic equipment tend to release less energy thus contributing to less internal heat loads inside the space.

Modulation and Heat Dissipation Techniques

The modulation and heat dissipation techniques rely on natural heat sinks to store and remove the internal heat gains. Examples of natural sinks are night sky, earth soil, and building mass. Therefore passive cooling techniques that use heat sinks can act to either modulate heat gain with thermal mass or dissipate heat through natural cooling strategies.

- Thermal mass - Heat gain modulation of an indoor space can be achieved by the proper use of the building's thermal mass as a heat sink. The thermal mass will absorb and store heat during daytime hours and return it to the space at a later time. Thermal mass can be coupled with night ventilation natural cooling strategy if the stored heat that will be delivered to the space during the evening/ night is not desirable.

- Natural cooling - Natural cooling refers to the use of ventilation or natural heat sinks for heat dissipation from indoor spaces. Natural cooling can be separated into five categories different categories: ventilation, night flushing, radiative cooling, evaporative cooling, and earth coupling.

Ventilation

Ventilation as a natural cooling strategy uses the physical properties of air to remove heat or provide cooling to occupants. In select cases, ventilation can be used to cool the building structure, which subsequently may serve as a heat sink.

- Cross ventilation - The strategy of cross ventilation relies on wind to pass through the building for the purpose of cooling the occupants. Cross ventilation requires openings on two sides of the space, called the inlet and outlet. The sizing and placement of the ventilation inlets and outlets will determine the direction and velocity of cross ventilation through the building. Generally, an equal (or greater) area of outlet openings must also be provided to provide adequate cross ventilation.

- Stack ventilation - Cross ventilation is an effective cooling strategy, however, wind is an unreliable resource. Stack ventilation is an alternative design strategy that relies on the buoyancy of warm air to rise and exit through openings located at ceiling height. Cooler outside air replaces the rising warm air through carefully designed inlets placed near the floor.

One specific application of natural ventilation is night flushing.

Night Flushing

Night flushing (also known as night ventilation, night cooling, night purging, or nocturnal convective cooling) is a passive or semi-passive cooling strategy that requires increased air movement at night to cool the structural elements of a building. Unlike free cooling, which assists in chilling water, night flushing cools down the thermal mass. To execute night flushing, the building envelope typically stays closed during the day, causing excess heat gains to be stored in the building's thermal mass. The building structure acts as a sink through the day and absorbs heat gains from occupants, equipment, solar radiation, and conduction through walls, roofs, and ceilings. At night, when the outside air is cooler and not too humid, the envelope is opened, allowing cooler air to pass through the building so the stored heat can be dissipated by convection. This process reduces the temperature of the indoor air and of the building's thermal mass, allowing convective, conductive, and radiant cooling to take place during the day when the building is occupied. Night flushing is most effective in climates with a large diurnal swing, i.e. a large difference between the daily maximum and minimum outdoor temperature. For optimal performance, the nighttime outdoor air temperature should fall below the daytime comfort zone limits of 22 °C (72 °F) and 60% relative humidity. For the night flushing strategy to be effective at reducing indoor temperature and energy usage, the thermal mass must be sized sufficiently and distributed over a wide enough surface area to absorb the space's daily heat gains. Also, the total air change rate must be high enough to remove the internal heat gains from the space at night. There are three ways night flushing can be achieved in a building:

- Natural night flushing by opening windows at night, letting wind-driven or buoyancy-driven airflow cool the space, and then closing windows during the day.
- Mechanical night flushing by forcing air mechanically through ventilation ducts at night at a high airflow rate and supplying air to the space during the day at a code-required minimum airflow rate.
- Mixed-mode night flushing through a combination of natural ventilation and mechanical ventilation, also known as mixed-mode ventilation, by using fans to assist the natural nighttime airflow.

There are numerous benefits to using night flushing as a cooling strategy for buildings, including improved comfort and a shift in peak energy load. Energy is most expensive during the day. By implementing night flushing, the usage of mechanical ventilation is reduced during the day, leading to energy and money savings.

There are also a number of limitations to using night flushing, such as usability, security, reduced indoor air quality, and poor room acoustics. For natural night flushing, the process of manually opening and closing windows every day can be tiresome, especially in the presence of insect screens. This problem can be eased with automated windows or ventilation louvers, such as in the Manitoba Hydro Place. Natural night flushing also requires windows to be open at night when the building is most likely unoccupied, which can raise security issues. If outdoor air is polluted, night flushing can expose occupants to harmful conditions inside the building. In loud city locations, the opening of windows can create poor acoustical conditions inside the building.

Radiative Cooling

All objects constantly emit and absorb radiant energy. An object will cool by radiation if the net flow is outward, which is the case during the night. At night, the long-wave radiation from the clear sky is less than the long-wave infrared radiation emitted from a building, thus there is a net flow to the sky. Since the roof provides the greatest surface visible to the night sky, designing the roof to act as a radiator is an effective strategy. There are two types of radiative cooling strategies that utilize the roof surface: direct and indirect:

- Direct radiant cooling - In a building designed to optimize direct radiation cooling, the building roof acts as a heat sink to absorb the daily internal loads. The roof acts as the best heat sink because it is the greatest surface exposed to the night sky. Radiate heat transfer with the night sky will remove heat from the building roof, thus cooling the building structure. Roof ponds are an example of this strategy. The roof pond design became popular with the development of the Sky thermal system designed by Harold Hay in 1977. There are various designs and configurations for the roof pond system but the concept is the same for all designs. The roof uses water, either plastic bags filled with water or an open pond, as the heat sink while a system of movable insulation panels regulate the mode of heating or cooling. During daytime in the summer, the water on the roof is protected from the solar radiation and ambient air temperature by movable insulation, which allows it to serve as a heat sink and absorb the heat generated inside through the ceiling. At night, the panels are retracted to allow nocturnal radiation between the roof pond and the night sky, thus removing the stored heat. In winter, the process is reversed so that the roof pond is allowed to absorb solar radiation during the day and release it during the night into the space below.
- Indirect radiant cooling - A heat transfer fluid removes heat from the building structure through radiate heat transfer with the night sky. A common design for this strategy involves a plenum between the building roof and the radiator surface. Air is drawn into the building through the plenum, cooled from the radiator, and cools the mass of the building structure. During the day, the building mass acts as a heat sink.

Evaporative Cooling

This design relies on the evaporative process of water to cool the incoming air while simultaneously increasing the relative humidity. A saturated filter is placed at the supply inlet so the natural process of evaporation can cool the supply air. Apart from the energy to drive the fans, water is the only other resource required to provide conditioning to indoor spaces. The effectiveness of evaporative cooling is largely dependent on the humidity of the outside air; dryer air produces more cooling. A study of field performance results in Kuwait revealed that power requirements for an evaporative cooler are approximately 75% less than the power requirements for a conventional packaged unit air-conditioner. As for interior comfort, a study found that evaporative cooling reduced inside air temperature by 9.6 °C compared to outdoor temperature.

Earth Coupling

Earth coupling uses the moderate and consistent temperature of the soil to act as a heat sink to cool a building through conduction. This passive cooling strategy is most effective when earth temperatures are cooler than ambient air temperature, such as in hot climates.

- Direct coupling or earth sheltering occurs when a building uses earth as a buffer for the walls. The earth acts as a heat sink and can effectively mitigate temperature extremes. Earth sheltering improves the performance of building envelopes by reducing heat losses and also reduces heat gains by limiting infiltration.
- Indirect coupling means that a building is coupled with the earth by means of earth ducts. An earth duct is a buried tube that acts as avenue for supply air to travel through before entering the building. The supply air is cooled by conductive heat transfer between the tubes and surrounding soil. Therefore, earth ducts will not perform well as a source of cooling unless the soil temperature is lower than the desired room air temperature. Earth ducts typically require long tubes to cool the supply air to an appropriate temperature before entering the building. A fan is required to draw the air from the earth duct into the building. Some of the factors that affect the performance of an earth duct are: duct length, number of bends, thickness of duct wall, depth of duct, diameter of the duct, and air velocity.

Energy Conservation Measure

An Energy conservation measure (ECM) is any type of project conducted, or technology implemented, to reduce the consumption of energy in a building. The types of projects implemented can be in a variety of forms but usually are designed to reduce utility costs: water, electricity and gas being the main three for industrial and commercial enterprises. The aim of an ECM should be to achieve a savings, reducing the amount of energy used by a particular process, technology or facility.

Energy conservation measures are often combined into larger guaranteed Energy Savings Performance Contracts to maximize energy savings while minimizing disruption to building occupants by coordinating renovations. Some ECMs cost less to implement yet return a higher energy savings. Traditionally, lighting projects were a good example of "low hanging fruit" that could be used to drive implementation of more substantial upgrades to HVAC systems in large facilities. Smaller buildings might combine window replacement with modern insulation using advanced building foams to improve energy performance. Energy dashboard projects are a new kind of ECM which relies on the behavioral change of building occupants to save energy. When implemented as part of a program, case studies (such as that for the DC Schools) report energy savings up 30%. Under the right circumstances, open energy dashboards can even be implemented for free to improve upon these savings even more.

On a global basis energy efficiency works behind the scenes to improve our energy security, lower our energy bills and move us closer to reaching our climate goals. According to the IEA, some 40% of the global energy efficiency market is financed with debt and equity. Energy Performance Investment are one financing mechanism by which ECMs can be implemented now and paid for by the savings realized over the life of the project. While all 50 states, Puerto Rico and Washington, D.C., have statutes allowing companies to offer energy savings performance contracts, success varies because of variations in the approach, the state's degree of involvement and other factors. Homes and businesses are implementing energy-efficiency measures that include low-energy lighting, insulation and even high tech energy dashboards to cut bills by avoiding waste and boosting productivity.

Businesses implementing ECMs in their commercial buildings often employ Energy Service Companies (ESCOs) experienced in energy performance contracting . This industry has been around since the 1970s and is more prevalent than ever today. The US-based organization EVO (Efficiency Valuation Organization) has created a set of guidelines for ESCOs to adhere to in evaluating the savings achieved by ECMs. These guidelines are called the International Performance Measurement and Verification Protocol (IPMVP).

Homeowners implementing ECMs in their residential buildings often start with an energy audit. This is a way homeowners look at what areas of their homes are using, and possibly losing energy. Residential energy auditors are accredited by the Building Performance Institute (BPI) or the Residential Energy Services Network (RESNET). Homeowners can hire a professional or do it themselves or use a smartphone to help do an audit.

Types of Energy Conservation Measures

Energy Dashboards

Energy Dashboards combine smart metering and Internet technologies to provide real-time data on energy use. Their success is based on the premise that real-time feedback drives behavior change and improves operational efficiency. Energy Dashboards are used to enable energy reduction competitions, showcase real-time building performance and green building features, and empower occupants to become active participants in energy management. Sustainability teams from major corporations, governments, universities and K–12 public schools use energy dashboards as an effective communication tool that creates transparency.

- BACnet + Drupal

- Building automation
- Gas meter
- Home Automation
- Smart grid and Smart meter
- Water metering

Insulation

Insulation decreases thermal losses in cold climates and thermal gains in hot climates thus reducing HVAC loads.

- House insulation
- Thermal insulation
- Cotton insulation
- Natural wool insulation
- VOCs in fiberglass insulation
- Cellulose insulation

Lighting

One of the simplest ways consumers save a copious amount of energy is switching incandescent light bulb to a compact fluorescent lamp (CFL). A 15W CFL is capable of providing just as much light as a 60W incandescent, while consuming just one fourth of the amount of energy.

- Compact fluorescent lamp
- Fluorescent bulbs
- LED lighting
- Linear fluorescent retrofit (T12 to T8)
- Sky lights
- Smart windows
- Solar charged flashlight
- Solar lights

Water

The average US homes wastes thousands of gallons of water a year. There are many water saving solutions that also save energy.

- 1.6 GPM or less low flow showerheads
- Ultra low flush toilet
- Composting toilets
- Faucet aerator

Windows

Windows may be one of the biggest contributing factors to energy loss and uncomfortable spaces. Individuals might find some ECMs related to windows more cost effective than others such as thermal curtains, films, or Smart windows.

References

- Yan Ji and Stellios Plainiotis (2006): Design for Sustainability. Beijing: China Architecture and Building Press. ISBN 7-112-08390-7.
- Hill, Steven, 2010, Europe's Promise: Why the European Way Is the Best Hope in an Insecure Age, University of California Press, ISBN 9780520248571.
- Frej, Anne, editor (2005). Green Office Buildings: A Practical Guide to Development. Urban Land Institute. pp. 138–142. ISBN 978-0-87420-937-2.
- Goyal, K. K. (July 14, 2009). Renewable Energy. New Delhi: Mahaveer & Sons. p. Influential Zero-and Low-Energy Buildings. ISBN 978-81-8377-243-3. Retrieved 18 September 2015.
- Givoni, Baruch (1994). Passive and Low Energy Cooling of Buildings (1st ed.). 605 Third Avenue, New York, NY 10158-0012, USA: John Wiley & Sons, Inc. ISBN 0-471-28473-4.
- Brown, G.Z.; DeKay, Mark (2001). Sun, wind, and light: architectural design strategies (2nd ed.). 605 Third Avenue, New York, NY 10158-0012, USA: John Wiley & Sons, Inc. ISBN 0-471-34877-5.
- Grondzik, Walter; Kwok, Alison; Stein, Benjamin; Reynolds, John (January 2011). Mechanical and Electrical Equipment for Buildings. John Wiley & Sons. ISBN 978-1-118-03940-3.
- DeKay, Mark; Brown, Charlie (December 2013). Sun, Wind, and Light: Architectural Design Strategies. John Wiley & Sons. ISBN 978-1-118-33288-7.
- "Renewable Energy Has a Variability Problem". The Green Economy Post: Green Careers, Green Business, Sustainability. 2010-11-30. Retrieved 2016-03-04.

6

Applications of Energy Conservation

The applications of energy conservation are renewable heat, smart grid, thermal energy storage and home energy saver. Renewable heat is a utilization of renewable energy. The heat technologies involved are solar heating, geothermal heating, heat pumps and heat exchangers. The aspects elucidated in the chapter are of vital importance, and provide a better understanding of energy conservation.

Renewable Heat

Renewable heat is an application of renewable energy and it refers to the renewable generation of heat, rather than electrical power (e.g. replacing a fossil fuel boiler using concentrating solar thermal to feed radiators). Renewable heat technologies include renewable biofuels, solar heating, geothermal heating, heat pumps and heat exchangers to recover lost heat. Significant attention is also applied to insulation.

Many colder countries consume more energy for heating than electrical power. For example, in 2005 the United Kingdom consumed 354 TWh of electric power, but had a heat requirement of 907 TWh, the majority of which (81%) was met using gas. The residential sector alone consumed a massive 550 TWh of energy for heating, mainly in the form of gas. Almost half of the final energy consumed in the UK (49%) was in the form of heat, of which 70% was used by households and in commercial and public buildings. Households used heat for mainly for space heating (69%) and heating water.

The relative competitiveness of renewable electricity and renewable heat depends on a nation's approach to energy and environment policy. Few renewable technologies (whether for heat, electricity or transport) are competitive with fossil fuels without some form of carbon valuation or subsidy. In those countries, such as Sweden, Denmark and Finland, where government intervention has been closest to a technology-neutral form of carbon valuation (i.e. carbon and energy taxes), renewable heat has played the leading role in a very substantial renewable contribution to final energy consumption. In those countries, such as Germany, Spain, the USA, and the UK, where government intervention has been set at different levels for different technologies, uses and scales, the contributions of renewable heat and renewable electricity technologies have depended on the relative levels of support, and have resulted generally in a lower renewable contribution to final energy consumption.

Leading Renewable Heat Technologies

Solar Heating

Solar heating is a style of building construction which uses the energy of summer or winter sunshine to provide an economic supply of primary or supplementary heat to a structure. The heat can be used for both space heating and water heating. Solar heating design is divided into two groups:

- Passive solar heating relies on the design and structure of the house to collect heat. Passive solar building design must also consider the storage and distribution of heat, which may be accomplished passively, or use air ducting to draw heat actively to the foundation of the building for storage. One such design was measured lifting the temperature of a house to 24 °C (75 °F) on a partially sunny winter day (-7 °C or 19 °F), and it is claimed that the system provides passively for the bulk of the building's heating. The 4,000-square-foot (370 m²) home cost $125 per square foot (or 370 m² at $1,351/m²), similar to the cost of a traditional new home.
- Active solar heating uses pumps to move air or a liquid from the solar collector into the building or storage area. Applications such as solar air heating and solar water heating typically capture solar heat in panels which can then be used for applications such as space heating and supplementation of residential water heaters. In contrast to photovoltaic panels, which are used to generate electricity, solar heating panels are less expensive and capture a much higher proportion of the sun's energy.

Solar heating systems usually require a small supplementary backup heating system, either conventional or renewable.

Geothermal Heating

Hot Springs located in Nevada.

Geothermal energy is accessed by drilling water or steam wells in a process similar to drilling for oil. Geothermal energy is an enormous, underused heat and power resource that is clean (emits little or no greenhouse gases), reliable (average system availability of 95%), and homegrown (making populations less dependent on oil).

The earth absorbs the sun's energy and stores it as heat in the oceans and underground. The ground temperature remains constant at a point of 42 to 100 °F (6 to 38 °C) all year round depending on where you live on earth. A geothermal heating system takes advantage of the consistent temperature found below the Earth's surface and uses it to heat and cool buildings. The system is made up of a series of pipes installed underground, connected to pipes in a building. A pump circulates liquid through the circuit. In the winter the fluid in the pipe absorbs the heat of the earth and uses it to heat the building. In the summer the fluid absorbs heat from the building and disposes of it in the earth.

Heat Pumps

Heat pumps use work to move heat from one place to another, and can be used for both heating and air conditioning. Though capital intensive, heat pumps are economical to run and can be powered by renewable electricity. Two common types of heat pump are air-source heat pumps (ASHP) and ground-source heat pumps (GSHP), depending on whether heat is transferred from the air or from the ground. Air source heat pumps are not effective when the outside air temperature is lower than about -15 °C, while ground-source heat pumps are not affected. The efficiency of a heat pump is measured by the coefficient of performance (CoP): For every unit of electricity used to pump the heat, an air source heat pump generates 2.5 to 3 units of heat (i.e. it has a CoP of 2.5 to 3), whereas a GSHP generates 3 to 3.5 units of heat. Based on current fuel prices for the United Kingdom, assuming a CoP of 3-4, a GSHP is sometimes a cheaper form of space heating than electric, oil, and solid fuel heating. Heat pumps can be linked to an interseasonal thermal energy storage (hot or cold), doubling the CoP from 4 to 8 by extracting heat from warmer ground.

Interseasonal Heat Transfer

A heat pump with Interseasonal Heat Transfer combines active solar collection to store surplus summer heat in thermal banks with ground-source heat pumps to extract it for space heating in winter. This reduces the "Lift" needed and doubles the CoP of the heat pump because the pump starts with warmth from the thermal bank in place of cold from the ground.

CoP and Lift

A heat pump CoP increases as the temperature difference, or "Lift", decreases between heat source and destination. The CoP can be maximized at design time by choosing a heating system requiring only a low final water temperature (e.g., underfloor heating), and by choosing a heat source with a high average temperature (e.g., the ground). Domestic hot water (DHW) and conventional radiators require high water temperatures, affecting the choice of heat pump technology. Low temperature radiators provide an alternative to conventional radiators.

Pump type and source	Typical use case	Heat Pump CoP variation with Output Temperature					
		35 °C (e.g. heated screed floor)	45 °C (e.g. low temp. radiator or heated screed floor)	55 °C (e.g. low temp. radiator or heated timber floor)	65 °C (e.g. std. radiator or DHW)	75 °C (e.g. std. radiator & DHW)	85 °C (e.g. std. radiator & DHW)
High Efficiency ASHP air at -20 °C		2.2	2.0	-	-	-	-
Two Stage ASHP air at -20 °C	Low source temp.	2.4	2.2	1.9	-	-	-
High Efficiency ASHP air at 0 °C	Low output temp.	3.8	2.8	2.2	2.0	-	-
Prototype Transcritical CO 2 (R744) Heat Pump with Tripartite Gas Cooler, source at 0 °C	High output temp.	3.3	-	-	4.2	-	3.0
GSHP water at 0 °C		5.0	3.7	2.9	2.4	-	-
GSHP ground at 10 °C	Low output temp.	7.2	5.0	3.7	2.9	2.4	-
Theoretical Carnot cycle limit, source -20 °C		5.6	4.9	4.4	4.0	3.7	3.4
Theoretical Carnot cycle limit, source 0 °C		8.8	7.1	6.0	5.2	4.6	4.2
Theoretical Lorentz Cycle limit (CO 2 pump), return fluid 25 °C, source 0 °C		10.1	8.8	7.9	7.1	6.5	6.1
Theoretical Carnot cycle limit, source 10 °C		12.3	9.1	7.3	6.1	5.4	4.8

Resistive Electrical Heating

Renewable electricity can be generated by hydropower, solar, wind, geothermal and by burning biomass. In a few countries where renewable electricity is inexpensive, resistance heating is common. In countries like Denmark where electricity is expensive, it is not permitted to install electric heating as the main heat source. Wind turbines have more output at night when there is a small demand for electricity, storage heaters consume this lower cost electricity at night and give off heat during the day.

Wood-pellet Heating

Wood Stove.

Wood Pellets.

Wood-pellet heating and other types of wood heating systems have achieved their greatest success in heating premises that are off the gas grid, typically being previously heated using heating oil or coal. Solid wood fuel requires a large amount of dedicated storage space, and the specialized heating systems can be expensive (though grant schemes are available in many European countries to offset this capital cost.) Low fuel costs mean that wood fuelled heating in Europe is frequently able to achieve a payback period of less than 3 to 5 years. Because of the large fuel storage requirement wood fuel can be less attractive in urban residential scenarios, or for premises connected to the gas grid (though rising gas prices and uncertainty of supply mean that wood fuel is becoming more competitive.) There is also growing concern over the air pollution from wood heating versus oil or gas heat, especially the fine particulates.

Wood-stove Heating

Burning wood fuel in an open fire is both extremely inefficient (0-20%) and polluting due to low temperature partial combustion. In the same way that a drafty building loses heat through loss of warm air through poor sealing, an open fire is responsible for large heat losses by drawing very large volumes of warm air out of the building.

Modern wood stove designs allow for more efficient combustion and then heat extraction. In the United States, new wood stoves are certified by the U.S. Environmental Protection Agency (EPA) and burn cleaner and more efficiently (the overall efficiency is 60-80%) and draw smaller volumes of warm air from the building.

"Cleaner" should not, however, be confused with clean. An Australian study of real-life emissions from woodheaters satisfying the current Australian standard, found that particle emissions averaged 9.4 g/kg wood burned (range 2.6 to 21.7). A heater with average wood consumption of 4 tonnes per year therefore emits 37.6 kg of PM2.5, i.e. particles less than 2.5 micrometers. This can be compared with a passenger car satisfying the current Euro 5 standards (introduced September 2009) of 0.005 g/km. So one new wood heater emits as much PM2.5 per year as 367 passenger cars each driving 20,000 km a year. A recent European study identified PM2.5 as the most health-hazardous air pollutant, causing an estimated 492,000 premature deaths. The next worst pollutant, ozone, is responsible for 21,000 premature deaths.

Because of the problems with pollution, the Australian Lung Foundation recommends using alternative means for climate control. The American Lung Association "strongly recommends using cleaner, less toxic sources of heat. Converting a wood-burning fireplace or stove to use either natural gas or propane will eliminate exposure to the dangerous toxins wood burning generates including dioxin, arsenic and formaldehyde.

"Renewable" should not be confused with "greenhouse neutral". A recent peer-reviewed paper found that, even if burning firewood from a sustainable supply, methane emissions from a typical Australian wood heater satisfying the current standard cause more global warming than heating the same house with gas. However, because a large proportion of firewood sold in the Australia is not from sustainable supplies, Australian households that use wood heating often cause more global warming than heating 3 similar homes with gas.

High efficiency stoves should meet the following design criteria:

- Well sealed and precisely calibrated to draw a low yet sufficient volume of air. Air-flow restriction is critical; a lower inflow of cold air cools the furnace less (a higher temperature is thus achieved). It also allows greater time for extraction of heat from the exhaust gas, and draws less heat from the building.
- The furnace must be well insulated to increase combustion temperature, and thus completeness.
- A well insulated furnace radiates little heat. Thus heat must be extracted instead from the exhaust gas duct. Heat absorption efficiencies are higher when the heat-exchange duct is longer, and when the flow of exhaust gas is slower.
- In many designs, the heat-exchange duct is built of a very large mass of heat-absorbing brick or stone. This design causes the absorbed heat to be emitted over a longer period - typically a day.

Renewable Natural Gas

Renewable natural gas is defined as gas obtained from biomass which is upgraded to a quality similar to natural gas.By upgrading the quality to that of natural gas, it becomes possible to distribute the gas to customers via the existing gas grid. According to the Energy research Centre of the Netherlands, renewable natural gas is 'cheaper than alternatives where biomass is used in a combined heat and power plant or local combustion plant'. Energy unit costs are lowered through 'favourable scale and operating hours', and end-user capital costs eliminated through distribution via the existing gas grid.

Energy Efficiency

Renewable heat goes hand in hand with energy efficiency. Indeed, renewable heating projects depend heavily for their success on energy efficiency; in the case of solar heating to cut reliance on the requirement supplementary heating, in the case of wood fuel heating to cut the cost of wood purchased and volume stored, and in the case of heat pumps to reduce the size and investment in heat pump, heat sink and electricity costs.

Two main types of improvement can be made to a building's energy efficiency:

Insulation

Improvements to insulation can cut energy consumption greatly, making a space cheaper to heat and to cool. However existing housing can often be difficult or expensive to improve. Newer buildings can benefit from many of the techniques of superinsulation. Older buildings can benefit from several kinds of improvement:

- Solid wall insulation: A building with solid walls can benefit from internal or external insulation. External wall insulation involves adding decorative weather-proof insulating panels or other treatment to the outside of the wall. Alternatively, internal wall insulation can be applied using ready-made insulation/plaster board laminates, or other methods. Thicknesses of internal or external insulation typically range between 50 and 100 mm.
- Cavity wall insulation: A building with cavity walls can benefit from insulation pumped into the cavity. This form of insulation is very cost effective.
- Programmable thermostats allow heating and cooling of a room to be switched off depending the time, day of the week, and temperature. A bedroom, for example, does not need to be heated during the day, but a living room does not need to be heated during the night.
- Roof insulation
- Insulated windows and doors
- Draught proofing

Underfloor Heating

Underfloor heating may sometimes be more energy efficient than traditional methods of heating:

- Water circulates within the system at low temperatures (35 °C - 50 °C) making gas boilers, wood fired boilers, and heat pumps significantly more efficient.
- Rooms with underfloor heating are cooler near the ceiling, where heat is not required, but warmer underfoot, where comfort is most required.
- Traditional radiators are frequently positioned underneath poorly insulated windows, heating them unnecessarily.

Waste-water Heat Recovery

It is possible to recover significant amounts of heat from waste hot water via hot water heat recycling. Major consumption of hot water is sinks, showers, baths, dishwashers, and clothes washers. On average 30% of a property's domestic hot water is used for showering. Incoming fresh water is typically of a far lower temperature than the waste water from a shower. An inexpensive heat exchanger recovers up on average 40% of the heat that would normally be wasted, by warming incoming cold fresh water with heat from outgoing waste water.

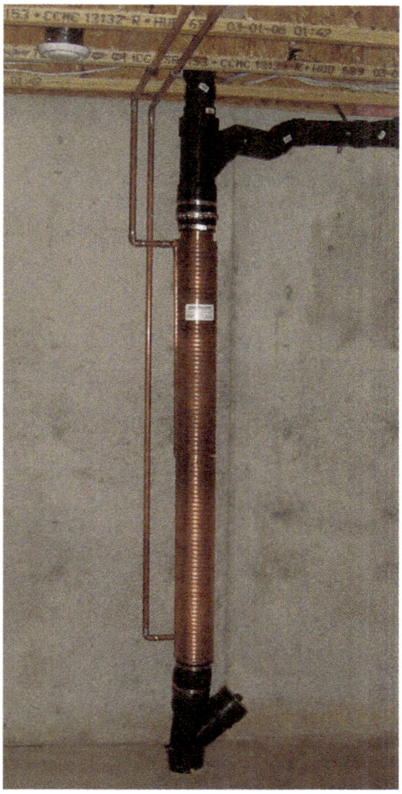

Recycling heat.

Smart Grid

A smart grid is an electrical grid which includes a variety of operational and energy measures including smart meters, smart appliances, renewable energy resources, and energy efficiency resources. Electronic power conditioning and control of the production and distribution of electricity are important aspects of the smart grid.

Smart grid policy is organized in Europe as Smart Grid European Technology Platform. Policy in the United States is described in 42 U.S.C. ch. 152, subch. IX § 17381.

Roll-out of smart grid technology also implies a fundamental re-engineering of the electricity services industry, although typical usage of the term is focused on the technical infrastructure.

Background

Historical development of the electricity grid

The first alternating current power grid system was installed in 1886 in Great Barrington, Massachusetts. At that time, the grid was a centralized unidirectional system of electric power transmission, electricity distribution, and demand-driven control.

In the 20th century local grids grew over time, and were eventually interconnected for economic and reliability reasons. By the 1960s, the electric grids of developed countries had become very large, mature and highly interconnected, with thousands of 'central' generation power stations delivering power to major load centres via high capacity power lines which were then branched and divided to provide power to smaller industrial and domestic users over the entire supply area. The topology of the 1960s grid was a result of the strong economies of scale: large coal-, gas- and oil-fired power stations in the 1 GW (1000 MW) to 3 GW scale are still found to be cost-effective, due to efficiency-boosting features that can be cost effective only when the stations become very large.

Power stations were located strategically to be close to fossil fuel reserves (either the mines or wells themselves, or else close to rail, road or port supply lines). Siting of hydro-electric dams in mountain areas also strongly influenced the structure of the emerging grid. Nuclear power plants were sited for availability of cooling water. Finally, fossil fuel-fired power stations were initially very polluting and were sited as far as economically possible from population centres once electricity distribution networks permitted it. By the late 1960s, the electricity grid reached the overwhelming majority of the population of developed countries, with only outlying regional areas remaining 'off-grid'.

Metering of electricity consumption was necessary on a per-user basis in order to allow appropriate billing according to the (highly variable) level of consumption of different users. Because of limited data collection and processing capability during the period of growth of the grid, fixed-tariff arrangements were commonly put in place, as well as dual-tariff arrangements where night-time power was charged at a lower rate than daytime power. The motivation for dual-tariff arrangements was the lower night-time demand. Dual tariffs made possible the use of low-cost night-time electrical power in applications such as the maintaining of 'heat banks' which served to 'smooth out' the daily demand, and reduce the number of turbines that needed to be turned off overnight, thereby improving the utilisation and profitability of the generation and transmission facilities. The metering capabilities of the 1960s grid meant technological limitations on the degree to which price signals could be propagated through the system.

Through the 1970s to the 1990s, growing demand led to increasing numbers of power stations. In some areas, supply of electricity, especially at peak times, could not keep up with this demand, resulting in poor power quality including blackouts, power cuts, and brownouts. Increasingly, electricity was depended on for industry, heating, communication, lighting, and entertainment, and consumers demanded ever higher levels of reliability.

Towards the end of the 20th century, electricity demand patterns were established:

domestic heating and air-conditioning led to daily peaks in demand that were met by an array of 'peaking power generators' that would only be turned on for short periods each day. The relatively low utilisation of these peaking generators (commonly, gas turbines were used due to their relatively lower capital cost and faster start-up times), together with the necessary redundancy in the electricity grid, resulted in high costs to the electricity companies, which were passed on in the form of increased tariffs. In the 21st century, some developing countries like China, India and Brazil were seen as pioneers of smart grid deployment.

Modernization Opportunities

Since the early 21st century, opportunities to take advantage of improvements in electronic communication technology to resolve the limitations and costs of the electrical grid have become apparent. Technological limitations on metering no longer force peak power prices to be averaged out and passed on to all consumers equally. In parallel, growing concerns over environmental damage from fossil-fired power stations has led to a desire to use large amounts of renewable energy. Dominant forms such as wind power and solar power are highly variable, and so the need for more sophisticated control systems became apparent, to facilitate the connection of sources to the otherwise highly controllable grid. Power from photovoltaic cells (and to a lesser extent wind turbines) has also, significantly, called into question the imperative for large, centralised power stations. The rapidly falling costs point to a major change from the centralised grid topology to one that is highly distributed, with power being both generated *and* consumed right at the limits of the grid. Finally, growing concern over terrorist attack in some countries has led to calls for a more robust energy grid that is less dependent on centralised power stations that were perceived to be potential attack targets.

Definition of "Smart Grid"

The first official definition of Smart Grid was provided by the Energy Independence and Security Act of 2007 (EISA-2007), which was approved by the US Congress in January 2007, and signed to law by President George W. Bush in December 2007. Title XIII of this bill provides a description, with ten characteristics, that can be considered a definition for Smart Grid, as follows:

"It is the policy of the United States to support the modernization of the Nation's electricity transmission and distribution system to maintain a reliable and secure electricity infrastructure that can meet future demand growth and to achieve each of the following, which together characterize a Smart Grid: (1) Increased use of digital information and controls technology to improve reliability, security, and efficiency of the electric grid. (2) Dynamic optimization of grid operations and resources, with full cyber-security. (3) Deployment and integration of distributed resources and generation, including renewable resources. (4) Development and incorporation of demand response, demand-side resources, and energy-efficiency resources. (5) Deployment of `smart' technologies (real-time, automated, interactive technologies that optimize the physical operation of appliances and consumer devices) for metering, communications concerning grid operations and status, and distribution automation. (6) Integration of `smart' appliances and consumer devices. (7) Deployment and integration of advanced electricity storage and peak-shaving technologies, including plug-in electric and hybrid electric vehicles, and thermal storage air conditioning. (8) Provision to consumers of timely information and control options. (9) Development of standards for communication and interoperability of appliances and equipment connected to the electric grid, including the infrastructure serving the grid. (10) Identification and lowering of unreasonable or unnecessary barriers to adoption of smart grid technologies, practices, and services."

A common element to most definitions is the application of digital processing and communications to the power grid, making data flow and information management central to the smart grid. Various capabilities result from the deeply integrated use of digital technology with power grids. Integration of the new grid information is one of the key issues in the design of smart grids. Electric utilities now find themselves making three classes of transformations: improvement of infrastructure, called the *strong grid* in China; addition of the digital layer, which is the essence of the *smart grid*; and business process transformation, necessary to capitalize on the investments in smart technology. Much of the work that has been going on in electric grid modernization, especially substation and distribution automation, is now included in the general concept of the smart grid.

Early Technological Innovations

Smart grid technologies emerged from earlier attempts at using electronic control, metering, and monitoring. In the 1980s, automatic meter reading was used for monitoring loads from large customers, and evolved into the Advanced Metering Infrastructure of the 1990s, whose meters could store how electricity was used at different times of the day. Smart meters add continuous communications so that monitoring can be done in real time, and can be used as a gateway to demand response-aware devices and "smart sockets" in the home. Early forms of such demand side management technologies were dynamic demand aware devices that passively sensed the load on the grid by monitoring changes in the power supply frequency. Devices such as industrial and domestic air conditioners, refrigerators and heaters adjusted their duty cycle to avoid activation during times the grid was suffering a peak condition. Beginning in 2000, Italy's Telegestore Project was the first to network large numbers (27 million) of homes using smart meters connected via low bandwidth power line communication. Some experiments used the term broadband over power lines (BPL), while others used wireless technologies such as mesh networking promoted for more reliable connections to disparate devices in the home as well as supporting metering of other utilities such as gas and water.

Monitoring and synchronization of wide area networks were revolutionized in the early

1990s when the Bonneville Power Administration expanded its smart grid research with prototype sensors that are capable of very rapid analysis of anomalies in electricity quality over very large geographic areas. The culmination of this work was the first operational Wide Area Measurement System (WAMS) in 2000. Other countries are rapidly integrating this technology — China started having a comprehensive national WAMS system when the past 5-year economic plan completed in 2012.

The earliest deployments of smart grids include the Italian system *Telegestore* (2005), the mesh network of Austin, Texas (since 2003), and the smart grid in Boulder, Colora-do (2008).

Features of the Smart Grid

The smart grid represents the full suite of current and proposed responses to the challenges of electricity supply. Because of the diverse range of factors there are numerous competing taxonomies and no agreement on a universal definition. Nevertheless, one possible categorisation is given here.

Reliability

The smart grid will make use of technologies, such as state estimation, that improve fault detection and allow self-healing of the network without the intervention of technicians. This will ensure more reliable supply of electricity, and reduced vulnerability to natural disasters or attack.

Although multiple routes are touted as a feature of the smart grid, the old grid also featured multiple routes. Initial power lines in the grid were built using a radial model, later connectivity was guaranteed via multiple routes, referred to as a network structure. However, this created a new problem: if the current flow or related effects across the network exceed the limits of any particular network element, it could fail, and the current would be shunted to other network elements, which eventually may fail also, causing a domino effect. A technique to prevent this is load shedding by rolling blackout or voltage reduction (brownout).

The economic impact of improved grid reliability and resilience is the subject of a number of studies and can be calculated using a US DOE funded methodology for US locations using at least one calculation tool.

Flexibility in Network Topology

Next-generation transmission and distribution infrastructure will be better able to handle possible bidirection energy flows, allowing for distributed generation such as from photovoltaic panels on building roofs, but also the use of fuel cells, charging to/from the batteries of electric cars, wind turbines, pumped hydroelectric power, and other sources.

Classic grids were designed for one-way flow of electricity, but if a local sub-network generates more power than it is consuming, the reverse flow can raise safety and reliability issues. A smart grid aims to manage these situations.

Efficiency

Numerous contributions to overall improvement of the efficiency of energy infrastructure are anticipated from the deployment of smart grid technology, in particular including demand-side management, for example turning off air conditioners during short-term spikes in electricity price, reducing the voltage when possible on distribution lines through Voltage/VAR Optimization (VVO), eliminating truck-rolls for meter reading, and reducing truck-rolls by improved outage management using data from Advanced Metering Infrastructure systems. The overall effect is less redundancy in transmission and distribution lines, and greater utilization of generators, leading to lower power prices.

Load Adjustment/Load balancing

The total load connected to the power grid can vary significantly over time. Although the total load is the sum of many individual choices of the clients, the overall load is not a stable, slow varying, increment of the load if a popular television program starts and millions of televisions will draw current instantly. Traditionally, to respond to a rapid increase in power consumption, faster than the start-up time of a large generator, some spare generators are put on a dissipative standby mode. A smart grid may warn all individual television sets, or another larger customer, to reduce the load temporarily (to allow time to start up a larger generator) or continuously (in the case of limited resources). Using mathematical prediction algorithms it is possible to predict how many standby generators need to be used, to reach a certain failure rate. In the traditional grid, the failure rate can only be reduced at the cost of more standby generators. In a smart grid, the load reduction by even a small portion of the clients may eliminate the problem.

Peak Curtailment/Leveling and Time of Use Pricing

To reduce demand during the high cost peak usage periods, communications and metering technologies inform smart devices in the home and business when energy demand is high and track how much electricity is used and when it is used. It also gives utility companies the ability to reduce consumption by communicating to devices directly in order to prevent system overloads. Examples would be a utility reducing the usage of a group of electric vehicle charging stations or shifting temperature set points of air conditioners in a city. To motivate them to cut back use and perform what is called peak curtailment or peak leveling, prices of electricity are increased during high demand periods, and decreased during low demand periods. It is thought that consumers and businesses will tend to consume less during high demand periods if it is possible for consumers and consumer devices to be aware of the high price premium for using electricity at peak periods. This could mean making trade-offs such as cycling on/off air conditioners or running dishwashers at 9 pm instead of 5 pm. When businesses and consumers see a direct economic benefit of using energy at off-peak times, the theory is that they will include energy cost of operation into their consumer device and building construction decisions and hence become more energy efficient.

According to proponents of smart grid plans,this will reduce the amount of spinning reserve that atomic utilities have to keep on stand-by, as the load curve will level itself through a combination of "invisible hand" free-market capitalism and central control of a large number of devices by power management services that pay consumers a portion of the peak power saved by turning their device off.

Sustainability

The improved flexibility of the smart grid permits greater penetration of highly variable renewable energy sources such as solar power and wind power, even without the addition of energy storage. Current network infrastructure is not built to allow for many distributed feed-in points, and typically even if some feed-in is allowed at the local (distribution) level, the transmission-level infrastructure cannot accommodate it. Rapid fluctuations in distributed generation, such as due to cloudy or gusty weather, present significant challenges to power engineers who need to ensure stable power levels through varying the output of the more controllable generators such as gas turbines and hydroelectric generators. Smart grid technology is a necessary condition for very large amounts of renewable electricity on the grid for this reason.

Market-enabling

The smart grid allows for systematic communication between suppliers (their energy price) and consumers (their willingness-to-pay), and permits both the suppliers and the consumers to be more flexible and sophisticated in their operational strategies. Only the critical loads will need to pay the peak energy prices, and consumers will be able to be more strategic in when they use energy. Generators with greater flexibility will be able to sell energy strategically for maximum profit, whereas inflexible generators such as base-load steam turbines and wind turbines will receive a varying tariff based on the level of demand and the status of the other generators currently operating. The overall effect is a signal that awards energy efficiency, and energy consumption that is sensitive to the time-varying limitations of the supply. At the domestic level, appliances with a degree of energy storage or thermal mass (such as refrigerators, heat banks, and heat pumps) will be well placed to 'play' the market and seek to minimise energy cost by adapting demand to the lower-cost energy support periods. This is an extension of the dual-tariff energy pricing mentioned above.

Demand Response Support

Demand response support allows generators and loads to interact in an automated fashion in real time, coordinating demand to flatten spikes. Eliminating the fraction of demand that occurs in these spikes eliminates the cost of adding reserve generators, cuts wear and tear and extends the life of equipment, and allows users to cut their energy bills by telling low priority devices to use energy only when it is cheapest.

Currently, power grid systems have varying degrees of communication within control systems for their high-value assets, such as in generating plants, transmission lines, substations and major energy users. In general information flows one way, from the users and the loads they control back to the utilities. The utilities attempt to meet the demand and succeed or fail to varying degrees (brownout, rolling blackout, uncontrolled blackout). The total amount of power demand by the users can have a very wide probability distribution which requires spare generating plants in standby mode to respond to the rapidly changing power usage. This one-way flow of information is expensive; the last 10% of generating capacity may be required as little as 1% of the time, and brownouts and outages can be costly to consumers.

Demand response can be provided by commercial, residential loads, and industrial loads. For example, Alcoa's Warrick Operation is participating in MISO as a qualified Demand Response Resource, and the Trimet Aluminium uses its smelter as a short-term mega-battery.

Latency of the data flow is a major concern, with some early smart meter architectures allowing actually as long as 24 hours delay in receiving the data, preventing any possible reaction by either supplying or demanding devices.

Platform for Advanced Services

As with other industries, use of robust two-way communications, advanced sensors, and distributed computing technology will improve the efficiency, reliability and safety of power delivery and use. It also opens up the potential for entirely new services or improvements on existing ones, such as fire monitoring and alarms that can shut off power, make phone calls to emergency services, etc.

Provision Megabits, Control Power With Kilobits, Sell the Rest

The amount of data required to perform monitoring and switching one's appliances off automatically is very small compared with that already reaching even remote homes to support voice, security, Internet and TV services. Many smart grid bandwidth upgrades are paid for by over-provisioning to also support consumer services, and subsidizing the communications with energy-related services or subsidizing the energy-related services, such as higher rates during peak hours, with communications. This is particularly true where governments run both sets of services as a public monopoly.

Because power and communications companies are generally separate commercial enterprises in North America and Europe, it has required considerable government and large-vendor effort to encourage various enterprises to cooperate. Some, like Cisco, see opportunity in providing devices to consumers very similar to those they have long been providing to industry. Others, such as Silver Spring Networks or Google, are data integrators rather than vendors of equipment. While the AC power control standards suggest powerline networking would be the primary means of communication among smart grid and home devices, the bits may not reach the home via Broadband over Power Lines (BPL) initially but by fixed wireless.

Technology

The bulk of smart grid technologies are already used in other applications such as manufacturing and telecommunications and are being adapted for use in grid operations.

- Integrated communications: Areas for improvement include: substation automation, demand response, distribution automation, supervisory control and data acquisition (SCADA), energy management systems, wireless mesh networks and other technologies, power-line carrier communications, and fiber-optics. Integrated communications will allow for real-time control, information and data exchange to optimize system reliability, asset utilization, and security.

- Sensing and measurement: core duties are evaluating congestion and grid stability, monitoring equipment health, energy theft prevention, and control strategies support. Technologies include: advanced microprocessor meters (smart meter) and meter reading equipment, wide-area monitoring systems, dynamic line rating (typically based on online readings by Distributed temperature sensing combined with Real time thermal rating (RTTR) systems), electromagnetic signature measurement/analysis, time-of-use and real-time pricing tools, advanced switches and cables, backscatter radio technology, and Digital protective relays.

- Smart meters.

- Phasor measurement units. Many in the power systems engineering community believe that the Northeast blackout of 2003 could have been contained to a much smaller area if a wide area phasor measurement network had been in place.

- Distributed power flow control: power flow control devices clamp onto existing transmission lines to control the flow of power within. Transmission lines enabled with such devices support greater use of renewable energy by providing more consistent, real-time control over how that energy is routed within the grid. This technology enables the grid to more effectively store intermittent energy from renewables for later use.

- Smart power generation using advanced components: smart power generation is a concept of matching electricity generation with demand using multiple identical generators which can start, stop and operate efficiently at chosen load, independently of the others, making them suitable for base load and peaking power generation. Matching supply and demand, called load balancing, is essential for a stable and reliable supply of electricity. Short-term deviations in the balance lead to frequency variations and a prolonged mismatch results in blackouts. Operators of power transmission systems are charged with the balancing task, matching the power output of all the generators to the load of their electrical grid. The load balancing task has become much more challenging as increasingly intermittent and variable generators such as wind turbines and solar cells are added to the grid, forcing other producers to adapt their output much more frequently than has been required in the past. First two dynamic grid stability power plants utilizing the concept has been ordered by Elering and will be built by Wärtsilä in Kiisa, Estonia (Kiisa Power Plant). Their purpose is to "provide dynamic generation capacity to meet sudden and unexpected drops in the electricity supply." They are scheduled to be ready during 2013 and 2014, and their total output will be 250 MW.
- Power system automation enables rapid diagnosis of and precise solutions to specific grid disruptions or outages. These technologies rely on and contribute to each of the other four key areas. Three technology categories for advanced control methods are: distributed intelligent agents (control systems), analytical tools (software algorithms and high-speed computers), and operational applications (SCADA, substation automation, demand response, etc.). Using artificial intelligence programming techniques, Fujian power grid in China created a wide area protection system that is rapidly able to accurately calculate a control strategy and execute it. The Voltage Stability Monitoring & Control (VSMC) software uses a sensitivity-based successive linear programming method to reliably determine the optimal control solution.

Research

Major Programs

IntelliGrid – Created by the Electric Power Research Institute (EPRI), IntelliGrid architecture provides methodology, tools, and recommendations for standards and technologies for utility use in planning, specifying, and procuring IT-based systems, such as advanced metering, distribution automation, and demand response. The architecture also provides a living laboratory for assessing devices, systems, and technology. Several utilities have applied IntelliGrid architecture including Southern California Edison, Long Island Power Authority, Salt River Project, and TXU Electric Delivery. The IntelliGrid Consortium is a public/private partnership that integrates and optimizes global research efforts, funds technology R&D, works to integrate technologies, and disseminates technical information.

Grid 2030 – Grid 2030 is a joint vision statement for the U.S. electrical system developed by the electric utility industry, equipment manufacturers, information technology providers, federal and state government agencies, interest groups, universities, and national laboratories. It covers generation, transmission, distribution, storage, and end-use. The National Electric Delivery Technologies Roadmap is the implementation document for the Grid 2030 vision. The Roadmap outlines the key issues and challenges for modernizing the grid and suggests paths that government and industry can take to build America's future electric delivery system.

Modern Grid Initiative (MGI) is a collaborative effort between the U.S. Department of Energy (DOE), the National Energy Technology Laboratory (NETL), utilities, consumers, researchers, and other grid stakeholders to modernize and integrate the U.S. electrical grid. DOE's Office of Electricity Delivery and Energy Reliability (OE) sponsors the initiative, which builds upon Grid 2030 and the National Electricity Delivery Technologies Roadmap and is aligned with other programs such as GridWise and GridWorks.

GridWise – A DOE OE program focused on developing information technology to modernize the U.S. electrical grid. Working with the GridWise Alliance, the program invests in communications architecture and standards; simulation and analysis tools; smart technologies; test beds and demonstration projects; and new regulatory, institutional, and market frameworks. The GridWise Alliance is a consortium of public and private electricity sector stakeholders, providing a forum for idea exchanges, cooperative efforts, and meetings with policy makers at federal and state levels.

GridWise Architecture Council (GWAC) was formed by the U.S. Department of Energy to promote and enable interoperability among the many entities that interact with the nation's electric power system. The GWAC members are a balanced and respected team representing the many constituencies of the electricity supply chain and users. The GWAC provides industry guidance and tools to articulate the goal of interoperability across the electric system, identify the concepts and architectures needed to make interoperability possible, and develop actionable steps to facilitate the inter operation of the systems, devices, and institutions that encompass the nation's electric system. The GridWise Architecture Council Interoperability Context Setting Framework, V 1.1 defines necessary guidelines and principles.

GridWorks – A DOE OE program focused on improving the reliability of the electric system through modernizing key grid components such as cables and conductors, substations and protective systems, and power electronics. The program›s focus includes coordinating efforts on high temperature superconducting systems, transmission reliability technologies, electric distribution technologies, energy storage devices, and GridWise systems.

Pacific Northwest Smart Grid Demonstration Project. - This project is a demonstration across five Pacific Northwest states-Idaho, Montana, Oregon, Washington, and Wyoming. It involves about 60,000 metered customers, and contains many key functions of the future smart grid.

Solar Cities - In Australia, the Solar Cities programme included close collaboration with energy companies to trial smart meters, peak and off-peak pricing, remote switching and related efforts. It also provided some limited funding for grid upgrades.

Smart Grid Modelling

Many different concepts have been used to model intelligent power grids. They are generally studied within the framework of complex systems. In a recent brainstorming session, the power grid was considered within the context of optimal control, ecology, human cognition, glassy dynamics, information theory, microphysics of clouds, and many others. Here is a selection of the types of analyses that have appeared in recent years.

Protection systems that verify and supervise themselves

Pelqim Spahiu and Ian R. Evans in their study introduced the concept of a substation based smart protection and hybrid Inspection Unit.

Kuramoto oscillators

The Kuramoto model is a well-studied system. The power grid has been described in this context as well. The goal is to keep the system in balance, or to maintain phase synchronization (also known as phase locking). Non-uniform oscillators also help to model different technologies, different types of power generators, patterns of consumption, and so on. The model has also been used to describe the synchronization patterns in the blinking of fireflies.

Bio-systems

Power grids have been related to complex biological systems in many other contexts. In one study, power grids were compared to the dolphin social network. These creatures streamline or intensify communication in case of an unusual situation. The intercommunications that enable them to survive are highly complex.

Random fuse networks

In percolation theory, random fuse networks have been studied. The current density might be too low in some areas, and too strong in others. The analysis can therefore be used to smooth out potential problems in the network. For instance, high-speed computer analysis can predict blown fuses and correct for them, or analyze patterns that might lead to a power outage. It is difficult for humans to predict the long term patterns in complex networks, so fuse or diode networks are used instead.

Neural networks

Neural networks have been considered for power grid management as well. Electric power systems can be classified in multiple different ways: non-linear, dynamic, discrete, random, and/or stochastic. Artificial Neural Networks (ANNs) attempt to solve the most difficult of these problems, the non-linear problems.

Demand Forecasting

One application of ANNs is in demand forecasting. In order for grids to operate economically and reliably, demand forecasting is essential, because it is used to predict the amount of power that will be consumed by the load. This is dependent on weather conditions, type of day, random events, incidents, etc. For non-linear loads though, the load profile isn't smooth and as predictable, resulting in higher uncertainty and less accuracy using the traditional Artificial Intelligence models. Some factors that ANNs consider when developing these sort of models: classification of load profiles of different customer classes based on the consumption of electricity, increased responsiveness of demand to predict real time electricity prices as compared to conventional grids, the need to input past demand as different components, such as peak load, base load, valley load, average load, etc. instead of joining them into a single input, and lastly, the dependence of the type on specific input variables. An example of the last case would be given the type of day, whether its weekday or weekend, that wouldn't have much of an effect on Hospital grids, but it'd be a big factor in resident housing grids' load profile.

Markov processes

As wind power continues to gain popularity, it becomes a necessary ingredient in realistic power grid studies. Off-line storage, wind variability, supply, demand, pricing, and other factors can be modelled as a mathematical game. Here the goal is to develop a winning strategy. Markov processes have been used to model and study this type of system.

Maximum entropy

All of these methods are, in one way or another, maximum entropy methods, which is an active area of research. This goes back to the ideas of Shannon, and many other researchers who studied communication networks. Continuing along similar lines today, modern wireless network research often considers the problem of network congestion, and many algorithms are being proposed to minimize it, including game theory, innovative combinations of FDMA, TDMA, and others.

Economics

Market Outlook

In 2009, the US smart grid industry was valued at about $21.4 billion – by 2014, it will exceed at least $42.8 billion. Given the success of the smart grids in the U.S., the world market is expected to grow at a faster rate, surging from $69.3 billion in 2009 to $171.4 billion by 2014. With the segments set to benefit the most will be smart metering hardware sellers and makers of software used to transmit and organize the massive amount of data collected by meters. Recently, the World Economic Forum reported a transformational investment of more than $7.6 trillion is needed over the next 25 years (or $300 billion per year) to modernize, expand, and decentralize the electricity infrastructure with technical innovation as key to the transformation.

General Economics Developments

As customers can choose their electricity suppliers, depending on their different tariff methods, the focus of transportation costs will be increased. Reduction of maintenance and replacements costs will stimulate more advanced control.

A smart grid precisely limits electrical power down to the residential level, network small-scale distributed energy generation and storage devices, communicate information on operating status and needs, collect information on prices and grid conditions, and move the grid beyond central control to a collaborative network.

US and UK Savings Estimates and Concerns

One United States Department of Energy study calculated that internal modernization of US grids with smart grid capabilities would save between 46 and 117 billion dollars over the next 20 years. As well as these industrial modernization benefits, smart grid features could expand energy efficiency beyond the grid into the home by coordinating low priority home devices such as water heaters so that their use of power takes advantage of the most desirable energy sources. Smart grids can also coordinate the production of power from large numbers of small power producers such as owners of rooftop solar panels — an arrangement that would otherwise prove problematic for power systems operators at local utilities.

One important question is whether consumers will act in response to market signals. The U.S. Department of Energy (DOE) as part of the American Recovery and Reinvestment Act Smart Grid Investment Grant and Demonstrations Program funded special consumer behavior studies to examine the acceptance, retention, and response of consumers subscribed to time-based utility rate programs that involve advanced metering infrastructure and customer systems such as in-home displays and programmable communicating thermostats.

Another concern is that the cost of telecommunications to fully support smart grids may be prohibitive. A less expensive communication mechanism is proposed using a form of "dynamic demand management" where devices shave peaks by shifting their loads in reaction to grid frequency. Grid frequency could be used to communicate load information without the need of an additional telecommunication network, but it would not support economic bargaining or quantification of contributions.

Although there are specific and proven smart grid technologies in use, *smart grid* is an aggregate term for a set of related technologies on which a specification is generally agreed, rather than a name for a specific technology. Some of the benefits of such a modernized electricity network include the ability to reduce power consumption at the consumer side during peak hours, called demand side management; enabling grid connection of distributed generation power (with photovoltaic arrays, small wind turbines, micro hydro, or even combined heat power generators in buildings); incorporating grid energy storage for distributed generation load balancing; and eliminating or containing failures such as widespread power grid cascading failures. The increased efficiency and reliability of the smart grid is expected to save consumers money and help reduce CO_2 emissions.

Oppositions and Concerns

Most opposition and concerns have centered on smart meters and the items (such as remote control, remote disconnect, and variable rate pricing) enabled by them. Where opposition to smart meters is encountered, they are often marketed as "smart grid" which connects smart grid to smart meters in the eyes of opponents. Specific points of opposition or concern include:

- consumer concerns over privacy, e.g. use of usage data by law enforcement
- social concerns over "fair" availability of electricity
- concern that complex rate systems (e.g. variable rates) remove clarity and accountability, allowing the supplier to take advantage of the customer
- concern over remotely controllable "kill switch" incorporated into most smart meters
- social concerns over Enron style abuses of information leverage
- concerns over giving the government mechanisms to control the use of all power using activities
- concerns over RF emissions from smart meters

Security

With the advent of cybercrime there is also concern on the security of the infrastructure, primarily that involving communications technology. Concerns chiefly center around the communications technology at the heart of the smart grid. Designed to allow real-time contact between utilities and meters in customers' homes and businesses, there is a risk that these capabilities could be exploited for criminal or even terrorist actions. One of the key capabilities of this connectivity is the ability to remotely switch off power supplies, enabling utilities to quickly and easily cease or modify supplies to customers who default on payment. This is undoubtedly a massive boon for energy providers, but also raises some significant security issues. Cybercriminals have infiltrated the U.S. electric grid before on numerous occasions. Aside from computer infiltration, there are also concerns that computer malware like Stuxnet, which targeted SCADA systems which are widely used in industry, could be used to attack a smart grid network.

Electricity theft is a concern in the U.S. where the smart meters being deployed use the RF technology of Fastrak transponders to communicate with the electricity transmission network. People with knowledge of electronics can devise interference devices to cause the smart meter to report lower than actual usage. Similarly, the same technology can be employed to make it appear that the energy the consumer is using is being used by another customer, increasing their bill.

Other Challenges to Adoption

Before a utility installs an advanced metering system, or any type of smart system, it must make a business case for the investment. Some components, like the power system stabilizers (PSS) installed on generators are very expensive, require complex integration in the grid's control system, are needed only during emergencies, and are only effective if other suppliers on the network have them. Without any incentive to install them, power suppliers don't. Most utilities find it difficult to justify installing a communications infrastructure for a single application (e.g. meter reading). Because of this, a utility must typically identify several applications that will use the same communications infrastructure – for example, reading a meter, monitoring power quality, remote connection and disconnection of customers, enabling demand response, etc. Ideally, the communications infrastructure will not only support near-term applications, but unanticipated applications that will arise in the future. Regulatory or legislative actions can also drive utilities to implement pieces of a smart grid puzzle. Each utility has a unique set of business, regulatory, and legislative drivers that guide its investments. This means that each utility will take a different path to creating their smart grid and that different utilities will create smart grids at different adoption rates.

Some features of smart grids draw opposition from industries that currently are, or hope to provide similar services. An example is competition with cable and DSL Internet providers from broadband over powerline internet access. Providers of SCADA control systems for grids have intentionally designed proprietary hardware, protocols and software so that they cannot inter-operate with other systems in order to tie its customers to the vendor.

Power Theft / Power Loss

Various "smart grid" systems have dual functions. This includes Advanced Metering Infrastructure systems which, when used with various software can be used to detect power theft and by process of elimination, detect where equipment failures have taken place. These are in addition to their primary functions of eliminating the need for human meter reading and measuring the time-of-use of electricity.

The worldwide power loss including theft is estimated at approximately two-hundred billion dollars annually.

Electricity theft also represents a major challenge when providing reliable electrical service in developing countries.

Deployments and Attempted Deployments

Enel. The earliest, and one of the largest, example of a smart grid is the Italian system installed by Enel S.p.A. of Italy. Completed in 2005, the Telegestore project was highly unusual in the utility world because the company designed and manufactured their own meters, acted as their own system integrator, and developed their own system software. The Telegestore project is widely regarded as the first commercial scale use of smart grid technology to the home, and delivers annual savings of 500 million euro at a project cost of 2.1 billion euro.

US Dept. of Energy - ARRA Smart Grid Project: One of the largest deployment programs in the world to-date is the U.S. Dept. of Energy's Smart Grid Program funded by the American Recovery and Reinvestment Act of 2009. This program required matching funding from individual utilities. A total of over $9 billion in Public/Private funds were invested as part of this program. Technologies included Advanced Metering Infrastructure, including over 65 million Advanced "Smart" Meters, Customer Interface Systems, Distribution & Substation Automation, Volt/VAR Optimization Systems, over 1,000 Synchrophasors, Dynamic Line Rating, Cyber Security Projects, Advanced Distribution Management Systems, Energy Storage Systems, and Renewable Energy Integration Projects. This program consisted of Investment Grants (matching), Demonstration Projects, Consumer Acceptance Studies, and Workforce Education Programs. Reports from all individual utility programs as well as overall impact reports will be completed by the second quarter of 2015.

Austin, Texas. In the US, the city of Austin, Texas has been working on building its smart grid since 2003, when its utility first replaced 1/3 of its manual meters with smart meters that communicate via a wireless mesh network. It currently manages 200,000 devices real-time (smart meters, smart thermostats, and sensors across its service area), and expects to be supporting 500,000 devices real-time in 2009 servicing 1 million consumers and 43,000 businesses.

Boulder, Colorado completed the first phase of its smart grid project in August 2008. Both systems use the smart meter as a gateway to the home automation network (HAN) that controls smart sockets and devices. Some HAN designers favor decoupling control functions from the meter, out of concern of future mismatches with new standards and technologies available from the fast moving business segment of home electronic devices.

Hydro One, in Ontario, Canada is in the midst of a large-scale Smart Grid initiative, deploying a standards-compliant communications infrastructure from Trilliant. By the end of 2010, the system will serve 1.3 million customers in the province of Ontario. The initiative won the "Best AMR Initiative in North America" award from the Utility Planning Network.

The City of Mannheim in Germany is using realtime Broadband Powerline (BPL) communications in its Model City Mannheim "MoMa" project.

Adelaide in Australia also plans to implement a localised green Smart Grid electricity network in the Tonsley Park redevelopment.

Sydney also in Australia, in partnership with the Australian Government implemented the Smart Grid, Smart City program.

Évora. InovGrid is an innovative project in Évora, Portugal that aims to equip the electricity grid with information and devices to automate grid management, improve service quality, reduce operating costs, promote energy efficiency and environmental sustainability, and increase the penetration of renewable energies and electric vehicles. It will be possible to control and manage the state of the entire electricity distribution grid at any given instant, allowing suppliers and energy services companies to use this technological platform to offer consumers information and added-value energy products and services. This project to install an intelligent energy grid places Portugal and EDP at the cutting edge of technological innovation and service provision in Europe.

E-Energy - In the so-called E-Energy projects several German utilities are creating first nucleolus in six independent model regions. A technology competition identified this model regions to carry out research and development activities with the main objective to create an "Internet of Energy."

Massachusetts. One of the first attempted deployments of "smart grid" technologies in the United States was rejected in 2009 by electricity regulators in the Commonwealth of Massachusetts, a US state. According to an article in the Boston Globe, Northeast Utilities' Western Massachusetts Electric Co. subsidiary actually attempted to create a "smart grid" program using public subsidies that would switch low income customers from post-pay to pre-pay billing (using "smart cards") in addition to special hiked "premium" rates for electricity used above a predetermined amount. This plan was rejected by regulators as it "eroded important protections for low-income customers against shutoffs".

According to the Boston Globe, the plan "unfairly targeted low-income customers and circumvented Massachusetts laws meant to help struggling consumers keep the lights on". A spokesman for an environmental group supportive of smart grid plans and Western Massachusetts' Electric's aforementioned "smart grid" plan, in particular, stated "If used properly, smart grid technology has a lot of potential for reducing peak demand, which would allow us to shut down some of the oldest, dirtiest power plants... It's a tool."

The eEnergy Vermont consortium is a US statewide initiative in Vermont, funded in part through the American Recovery and Reinvestment Act of 2009, in which all of the electric utilities in the state have rapidly adopted a variety of Smart Grid technologies, including about 90% Advanced Metering Infrastructure deployment, and are presently evaluating a variety of dynamic rate structures.

In the Netherlands a large-scale project (>5000 connections, >20 partners) was initiated to demonstrate integrated smart grids technologies, services and business cases.

LIFE Factory Microgrid (LIFE13 ENV / ES / 000700) is a demonstrative project that is part of the LIFE+ 2013 program (European Commission), whose main objective is to demonstrate, through the implementation of a full-scale industrial smartgrid that microgrids can become one of the most suitable solutions for energy generation and management in factories that want to minimize their environmental impact.

OpenADR Implementations

Certain deployments utilize the OpenADR standard for load shedding and demand reduction during higher demand periods.

China

The smart grid market in China is estimated to be $22.3 billion with a projected growth to $61.4 billion by 2015. Honeywell is developing a demand response pilot and feasibility study for China with the State Grid Corp. of China using the OpenADR demand response standard. The State Grid Corp., the Chinese Academy of Science, and General Electric intend to work together to develop standards for China's smart grid rollout.

United Kingdom

The OpenADR standard was demonstrated in Bracknell, England, where *peak* use in commercial buildings was reduced by 45 percent. As a result of the pilot, the Scottish and Southern Energy (SSE) said it would connect up to 30 commercial and industrial buildings in Thames Valley, west of London, to a demand response program.

United States

In 2009, the US Department of Energy awarded an $11 million grant to Southern California Edison and Honeywell for a demand response program that automatically turns down energy use during peak hours for participating industrial customers. The Department of Energy awarded an $11.4 million grant to Honeywell to implement the program using the OpenADR standard.

Hawaiian Electric Co. (HECO) is implementing a two-year pilot project to test the ability of an ADR program to respond to the intermittence of wind power. Hawaii has a goal to obtain 70 percent of its power from renewable sources by 2030. HECO will give customers incentives for reducing power consumption within 10 minutes of a notice.

Guidelines, Standards and User Groups

Part of the IEEE Smart Grid Initiative, IEEE 2030.2 represents an extension of the work aimed at utility storage systems for transmission and distribution networks. The IEEE P2030 group expects to deliver early 2011 an overarching set of guidelines on smart grid interfaces. The new guidelines will cover areas including batteries and supercapacitors as well as flywheels. The group has also spun out a 2030.1 effort drafting guidelines for integrating electric vehicles into the smart grid.

IEC TC57 has created a family of international standards that can be used as part of the smart grid. These standards include IEC61850 which is an architecture for substation automation, and IEC 61970/61968 – the Common Information Model (CIM). The CIM provides for common semantics to be used for turning data into information.

OpenADR is an open-source smart grid communications standard used for demand response applications. It is typically used to send information and signals to cause electrical power-using devices to be turned off during periods of higher demand.

MultiSpeak has created a specification that supports distribution functionality of the smart grid. MultiSpeak has a robust set of integration definitions that supports nearly all of the software interfaces necessary for a distribution utility or for the distribution portion of a vertically integrated utility. MultiSpeak integration is defined using extensible markup language (XML) and web services.

The IEEE has created a standard to support synchrophasors – C37.118.

The UCA International User Group discusses and supports real world experience of the standards used in smart grids.

A utility task group within LonMark International deals with smart grid related issues.

There is a growing trend towards the use of TCP/IP technology as a common communication platform for smart meter applications, so that utilities can deploy multiple communication systems, while using IP technology as a common management platform.

IEEE P2030 is an IEEE project developing a “Draft Guide for Smart Grid Interoperability of Energy Technology and Information Technology Operation with the Electric Power System (EPS), and End-Use Applications and Loads".

NIST has included ITU-T G.hn as one of the "Standards Identified for Implementation" for the Smart Grid "for which it believed there was strong stakeholder consensus". G.hn is standard for high-speed communications over power lines, phone lines and coaxial cables.

OASIS EnergyInterop' – An OASIS technical committee developing XML standards for energy interoperation. Its starting point is the California OpenADR standard.

Under the Energy Independence and Security Act of 2007 (EISA), NIST is charged with overseeing the identification and selection of hundreds of standards that will be required to implement the Smart Grid in the U.S. These standards will be referred by NIST to the Federal Energy Regulatory Commission (FERC). This work has begun, and the first standards have already been selected for inclusion in NIST's Smart Grid catalog. However, some commentators have suggested that the benefits that could be realized from Smart Grid standardization could be threatened by a growing number of patents that cover Smart Grid architecture and technologies. If patents that cover standardized Smart Grid elements are not revealed until technology is broadly distributed throughout the network ("locked-in"), significant disruption could occur when patent holders seek to collect unanticipated rents from large segments of the market.

Thermal Energy Storage

Thermal energy storage (TES) is achieved with greatly differing technologies that collectively accommodate a wide range of needs. It allows excess thermal energy to be collected for later use, hours, days or many months later, at individual building, multiuser building, district, town or even regional scale depending on the specific technology. As examples: energy demand can be balanced between daytime and nighttime; summer heat from solar collectors can be stored interseasonally for use in winter; and cold obtained from winter air can be provided for summer air conditioning. Storage mediums include: water or ice-slush tanks ranging from small to massive, masses of native earth or bedrock accessed with heat exchangers in clusters of small-diameter boreholes (sometimes quite deep); deep aquifers contained between impermeable strata; shallow, lined pits filled with gravel and water and top-insulated; and eutectic, phase-change materials.

Other sources of thermal energy for storage include heat or cold produced with heat pumps from off-peak, lower cost electric power, a practice called peak shaving; heat from combined heat and power (CHP) power plants; heat produced by renewable electrical energy that exceeds grid demand and waste heat from industrial processes. Heat storage, both seasonal and short term, is considered an important means for cheaply balancing high shares of variable renewable electricity production and integration of electricity and heating sectors in energy systems almost or completely fed by renewable energy.

Solar Energy Storage

Most practical active solar heating systems provide storage from a few hours to a day's worth of energy collected. However, there are a growing number of facilities that use seasonal thermal energy storage (STES), enabling solar energy to be stored in summer for space heating use during winter. The Drake Landing Solar Community in Alberta, Canada has now achieved a year-round 97% solar heating fraction, a world record made possible only by incorporating STES.

Molten salt is a means of storing heat at a high temperature. This is a current commercial technology used in conjunction with concentrated solar power for later use in electricity generation, to allow solar power to provide electricity on a more continuous basis. These molten salts (Potassium nitrate, Calcium nitrate, Sodium nitrate, Lithium nitrate, etc.) have the property to absorb and store the heat energy that is released to the water, to transfer energy when needed. To improve the salt properties it must be mixed in a eutectic mixture.

The use of both latent heat and sensible heat are also possible with high temperature solar thermal input. Various eutectic mixtures of metals, such as Aluminium and Silicon (AlSi12) offer a high melting point suited to efficient steam generation, while high alumina cement-based materials offer good thermal storage capabilities

Economics

High peak loads drive the capital expenditures of the electricity generation industry. The industry meets these peak loads with low-efficiency peaking power plants, usually gas turbines, which have lower capital costs and, since the recent drop in natural gas prices, have low fuel costs as well. A kilowatt-hour of electricity consumed at night can be produced at much lower marginal cost. Utilities have begun to pass these lower costs to consumers, in the form of Time of Use (TOU) rates, or Real Time Pricing (RTP) Rates.

Heat Storage in Tanks or Rock Caverns

A steam accumulator consists of an insulated steel pressure tank containing hot water and steam under pressure. As a heat storage device, it is used to mediate heat production by a variable or steady source from a variable demand for heat. Steam accumulators may take on a significance for energy storage in solar thermal energy projects.

Large stores are widely used in Scandinavia to store heat for several days, to decouple heat and power production and to help meet peak demands. Interseasonal storage in caverns has been investigated and appears to be economical.

Heat Storage in hot Rocks, Concrete, Pebbles Etc

Water has one of the highest thermal capacities Heat capacity - 4.2 J/(cm^3·K) whereas concrete has about one third of that. On the other hand, concrete can be heated to much higher temperatures – 1200 °C by e.g. electrical heating and therefore has a much higher overall volumetric capacity. Thus in the example below, an insulated cube of about 2.8 m would appear to provide sufficient storage for a single house to meet 50% of heating demand. This could, in principle, be used to store surplus wind or PV heat due to the ability of electrical heating to reach high temperatures. At the neighborhood level, the Wiggenhausen-Süd solar development at Friedrichshafen has received international attention. This features a 12,000 m^3 (420,000 cu ft) reinforced concrete thermal store linked to 4,300 m^2 (46,000 sq ft) of solar collectors, which will supply the 570 houses with around 50% of their heating and hot water. Siemens builds a 36 MWh thermal storage near Hamburg with 600°C basalt and 1.5 MW electric output.

Electric Thermal Storage Heaters

These are commonplace in European homes and consist of high-density ceramic bricks or feolite blocks heated to a high temperature with electricity, and well insulated to release heat over a number of hours.

Ice-based Technology

Air conditioning can be provided more economically by using cheaper electricity at night to freeze water into ice, then using the cool of the ice in the afternoon to reduce the electricity needed to handle air conditioning demands. Thermal energy storage using ice makes use of the large heat of fusion of water. One metric ton of water, one cubic meter, can store 334 million joules (MJ) or 317,000 BTUs (93kWh or 26.4 ton-hours). In fact, ice was originally transported from mountains to cities for use as a coolant, and the original definition of a "ton" of cooling capacity (heat flow) was the heat to melt one ton of ice every 24 hours. This is the heat flow one would expect in a 3,000-square-foot (280 m^2) house in Boston in the summer. This definition has since been replaced by less archaic units: one ton HVAC capacity = 12,000 BTU/hour (~3.5 kW). Either way, an agreeably small storage facility can hold enough ice to cool a large building for a day or a week, whether that ice is produced by anhydrous ammonia chillers or hauled in by horse-drawn carts.

As such there are developing and developed applications where ice is produced during off-peak periods and used for cooling at later time.

In addition to using ice in cooling applications it is also being used in heat pump based heating systems. In these applications the phase change energy provides a very significant layer of thermal capacity that is near the bottom range of temperature that water source heat pumps can operate in. This allows the system to ride out the heaviest heating load conditions and extends the timeframe by which the source energy elements can contribute heat back into the system.

Cryogenic Energy Storage

This uses liquification of air or nitrogen as an energy store.

A pilot cryogenic energy system that uses liquid air as the energy store, and low-grade waste heat to drive the thermal re-expansion of the air, has been operating at a power station in Slough, UK since 2010.

Molten Salt Technology

Molten salt can be employed as a thermal energy storage method to retain thermal energy collected by a solar tower or solar trough of a concentrated solar power plant, so that it can be used to generate electricity in bad weather or at night. It was demonstrated in the Solar Two project from 1995-1999. The system is predicted to have an annual efficiency of 99%, a reference to the energy retained by storing heat before turning it into electricity, versus converting heat directly into electricity. The molten salt mixtures vary. The most extended mixture contains sodium nitrate, potassium nitrate and calcium nitrate. It is non-flammable and nontoxic, and has already been used in the chemical and metals industries as a heat-transport fluid, so experience with such systems exists in non-solar applications.

The salt melts at 131 °C (268 °F). It is kept liquid at 288 °C (550 °F) in an insulated "cold" storage tank. The liquid salt is pumped through panels in a solar collector where the focused sun heats it to 566 °C (1,051 °F). It is then sent to a hot storage tank. This is so well insulated that the thermal energy can be usefully stored for up to a week.

When electricity is needed, the hot salt is pumped to a conventional steam-generator to produce superheated steam for a turbine/generator as used in any conventional coal, oil, or nuclear power plant. A 100-megawatt turbine would need a tank of about 9.1 metres (30 ft) tall and 24 metres (79 ft) in diameter to drive it for four hours by this design.

Several parabolic trough power plants in Spain and solar power tower developer SolarReserve use this thermal energy storage concept. The Solana Generating Station in the U.S. has 6 hours of storage by molten salt.

Hot Silicon Technology

Solid or molten silicon offers much higher storage temperatures than salts with consequent greater capacity and efficiency. It is being researched as a possible more energy efficient storage technology. Silicon is able to store more than 1MWh of energy per cubic metre at 1400°C.

Pumped-heat Electricity Storage

In pumped-heat electricity storage (PHES), a reversible heat-pump system is used to store energy as a temperature difference between two heat stores.

Isentropic

One system currently being developed by the UK based company Isentropic operates as follows. It comprises two insulated containers; a hot vessel storing thermal energy at high temperature and high pressure, and a cold vessel storing thermal energy at low temperature and low pressure. Both vessels are filled with crushed rock or gravel which acts as the heat storage medium. The vessels are connected at top and bottom by pipes and the whole system is filled with the inert gas argon.

During the charging cycle the system uses surplus electricity to work as a heat pump. Argon at ambient temperature and pressure from the top of the cold store is compressed adiabatically to a pressure of 12 bar, heating it to around 500 °C (900 °F). The compressed gas is transferred to the top of the hot vessel where it percolates down through the gravel, transferring its heat to the rock and cooling to ambient temperature. The cooled, but still pressurized, gas emerging at the bottom of the vessel is then expanded (again adiabatically) back down to 1 bar, which lowers its temperature to -150 °C. The cold gas is then passed up through the cold vessel where it cools the rock while being warmed back to its initial condition.

The energy is recovered as electricity by reversing the cycle. The hot gas from the hot vessel is expanded to drive a generator and then supplied to the cold store. The cooled gas retrieved from the bottom of the cold store is compressed which heats the gas to ambient temperature. The gas is then transferred to the bottom of the hot vessel to be reheated.

The compression and expansion processes are provided by a specially designed reciprocating machine using sliding valves. Surplus heat generated by inefficiencies in the process is shed to the environment through heat exchangers during the discharging cycle.

The developer claims that a round trip efficiency of 72-80% is achievable. This compares to >80% achievable with pumped hydro energy storage.

Electrolysis of Water

It has also been suggested that an innovative method of 'storing' solar energy is to use electricity generated by Photovoltaic panels to electrolyse water to produce hydrogen fuel that can be stored and used to power cars and fuel cells renewably and cleanly.

Other Systems

Another proposed system uses turbomachinery and is capable of operating at much higher power levels.

Research

Storing energy in molecular bonds is being investigated. Energy densities equivalent to lithium-ion batteries have been achieved.

Home Energy Saver

Home Energy Saver is a set of on–line resources developed by the U.S. Department of Energy at the Lawrence Berkeley National Laboratory intended to help consumers and professional energy analysts, analyze, reduce, and manage home energy use.

The Home Energy Saver energy assessment tool allows consumers to conduct a do-it-yourself home energy audit and provides specific recommendations to help lower household energy consumption and utility costs. By entering a zip code, users get estimates for typical and efficient homes in their area. The estimates break down energy consumption by "end use". End uses reported by Home Energy Saver include: heating, cooling, water heating, major appliances, small appliances, and lighting.

The more details a user enters, (e.g., insulation levels, roofing, age of major equipment, how systems are used) the more customized the assessment results and energy efficiency recommendations become. The tailored reports allows consumers to drill into estimated cost of improvements, anticipated payback time, projected utility bill savings, and how much energy use and green house gas production will be reduced. Consumers can vary the energy efficiency assumptions and the upgrade costs, (e.g., replacing the default values with actual estimates from contractors) and recalculate the payback times and other details.

The Home Energy Saver website includes a section called LEARN which offers tips about energy savings, an explanation of the house-as-system energy efficiency approach, and other information to help people understand how energy is used in a home.

When launched in 1994, Home Energy Saver was the first and only online home energy calculator. Thereafter, 6 million people have used it to analyze their home energy use. Nearly 1 million people visit the site each year. In 2009, a second version of the tool, Home Energy Saver Professional, was launched. This advanced version provides a low cost, interactive energy simulation/assessments tool for contractors, building professionals, weatherization professionals, and building designers.

The Home Energy Simulation Model

The Home Energy Saver is built on DOE-2, a computer program for building heating and cooling energy analysis and design. DOE-2 performs a thermal load simulation that accounts for heating and cooling equipment and thermal distribution efficiencies, infiltration, and thermostat management. User-entered zip codes are mapped to one of about 300 unique "weather tapes" that impose a year's worth of local weather conditions on the home to determine heating and cooling needs.

Home Energy Saver extends DOE-2 in a number of ways to improve the simulation model. For example, when users enter their actual electricity tariffs, the predictive power of the model improves. Other methods are used to calculate the energy used by appliances, water heating, and lighting.

The public domain HES calculation methods and underlying data are clearly documented on the website. Other web-based tool developers are welcome to use this information at no cost, providing that the source is properly credited.

Energy Saving Recommendations

The Home Energy Saver enables users to quantify the benefits of improving the energy efficiency and comfort of homes in the following ways:

- No Cost Changes – No cost changes are modifications to the way energy is used, like lowering the hot water heater temperature, unplugging the second refrigerator that is running to cool just a few things, doing laundry with cool or cold water instead of hot, or programming the thermostat a bit lower. These changes don't cost anything, but they can save a substantial amount of energy over time.
- Low Cost Changes – Low cost changes include actions like changing out incandescent light bulbs for compact fluorescent lamps (CFLs) or LED bulbs, wrapping a hot water heater in an insulating blanket, or weatherizing a home by caulking or adding weather stripping. Low cost changes are typically Do-It-Yourself tasks that can improve the energy efficiency of a home dramatically.
- Deep Home Energy Upgrades or Retrofits – Upgrades can include actions ranging from replacing old inefficient appliances with new Energy Star appliances, adding insulation, or replacing major systems like heating equipment or the roof.

The energy improvement recommendations are drawn from the National Residential Energy Efficiency Measures Database.

Awards & Recognition

Each year, the R&D 100 Awards recognize the year's 100 most significant, innovative, newly introduced research and development advances. The awards are recognized in industry, government, and academia as proof that a product is one of the most innovative ideas of the year, nationally and internationally. Home Energy Saver and Hohm received an R&D 100 Award in 2010.

Home Energy Saver received the U.S. Department of Energy's "Energy 100" award as one of the best 100 scientific and technological accomplishments over DOE's 23-year lifetime. The discoveries were chosen based on their impact in saving consumers money and improving quality of life.

PC Magazine recognized Home Energy Saver in 2004 as one of the "Top 100 Undiscovered Websites.

MSN-Money rates Home Energy Saver among the "Best Sites for Free Government Help" including it in the list of "The 100 most Useful Sites on the Internet.

Licensing

Organizations who want to provide their customers tools to predict home energy consumption can license the Home Energy Saver Application Programming Interfaces (APIs).

Microsoft, the first organization to license the Home Energy Saver, uses it to drive Microsoft Hohm.

References

- Berger, Lars T.; Iniewski, Krzysztof, eds. (April 2012). Smart Grid - Applicacions, Communications and Security. John Wiley and Sons. ISBN 978-1-1180-0439-5.
- Klimstra, Jakob; Hotakainen, Markus (2011). Smart Power Generation (PDF). Helsinki: Avain Publishers. ISBN 9789516928466.
- Ehrlich, Robert, 2013, Renewable Energy: A First Course, CRC Press, Chap. 13.1.22 Thermal storage p. 375 ISBN 978-1439861158
- "Siemens project to test heated rocks for large-scale, low-cost thermal energy storage". Utility Dive. 12 October 2016. Retrieved 15 October 2016.
- "Assessment of the health impacts of exposure to PM2.5 at a European levelETC/ACC Technical Paper 2009/1". Air-climate.eionet.europa.eu. 2009-06-29. Retrieved 2013-10-02.
- "American Lung Association Cautions Against Wood-burning and Urges Cleaner Alternatives for Winter Heat - American Lung Association". Lungusa.org. 2008-09-29. Retrieved 2013-10-02.
- Gorman, Siobahn (6 April 2008). "Electricity Grid in U.S. Penetrated by Spies". Wall Street Journal. Retrieved 20 January 2012.
- Lundin, Barbara (January 24, 2012). "Honeywell builds on smart grid success in England". Fierce SmartGrid. Retrieved March 7, 2012.
- John, Jeff (February 2, 2012). "Balancing Hawaiian Wind Power with Demand Response". GreenTechMedia. Retrieved March 7, 2012.

7

Environmental Impact of the Energy Industry

Humans have used energy for hundreds of years. We have consumed natural resources to the extent that it has caused global warming. Changes in most parts of the worlds can already be noticed. The following section will not only provide an overview but it also delves into the topics related to it.

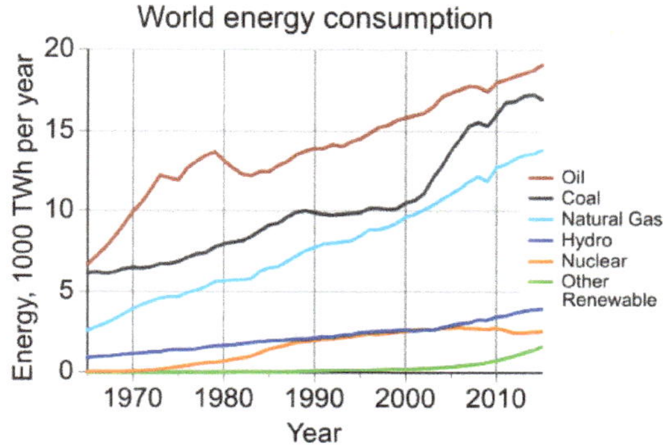

World consumption of primary energy by energy type.

The environmental impact of the energy industry is diverse. Energy has been harnessed by human beings for millennia. Initially it was with the use of fire for light, heat, cooking and for safety, and its use can be traced back at least 1.9 million years. In recent years there has been a trend towards the increased commercialization of various renewable energy sources.

Consumption of fossil fuel resources leads to global warming and climate change. In most parts of the world little change is being made to slow these changes. If the peak oil theory proves true, and more explorations of viable alternative energy sources are made, our impact could be less hostile to our environment.

Rapidly advancing technologies can achieve a transition of energy generation, water and waste management, and food production towards better environmental and energy usage practices using methods of systems ecology and industrial ecology.

Issues

Climate Change

The scientific consensus on global warming and climate change is that it is caused by anthropogenic greenhouse gas emissions, the majority of which comes from burning fossil fuels with deforestation and some agricultural practices being also major contributors.

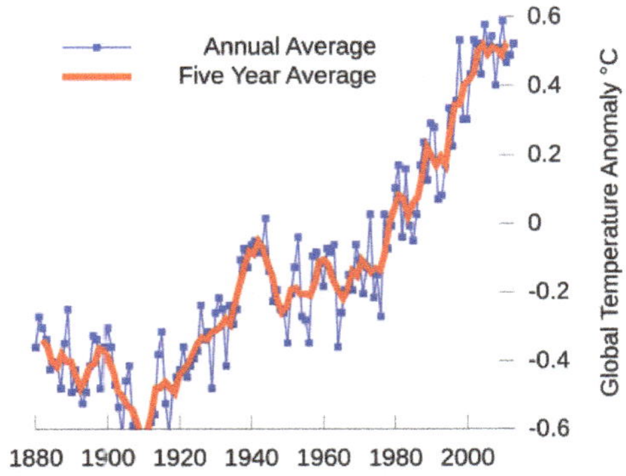

Global mean surface temperature anomaly relative to 1961–1990.

Although there is a highly publicized denial of climate change, the vast majority of scientists working in climatology accept that it is due to human activity. The IPCC report *Climate Change 2007: Climate Change Impacts, Adaptation and Vulnerability* predicts that climate change will cause shortages of food and water and increased risk of flooding that will affect billions of people, particularly those living in poverty.

One measurement of greenhouse gas related and other Externality comparisons between energy sources can be found in the ExternE project by the Paul Scherrer Institut and the University of Stuttgart which was funded by the European Commission. According to that study, hydroelectric electricity produces the lowest CO2 emissions, wind produces the second-lowest, nuclear energy produces the third-lowest and solar photovoltaic produces the fourth-lowest.

Similarly, the same research study (ExternE, Externalities of Energy), undertaken from 1995 to 2005 found that the cost of producing electricity from coal or oil would double over its present value, and the cost of electricity production from gas would increase by 30% if external costs such as damage to the environment and to human health, from the airborne particulate matter, nitrogen oxides, chromium VI and arsenic emissions produced by these sources, were taken into account. It was estimated in the study that these external, downstream, fossil fuel costs amount up to 1%-2% of the EU's entire Gross Domestic Product (GDP), and this was before the external cost of global warming from these sources was even included. The study also found that the environmental and health costs of nuclear power, per unit of energy delivered, was €0.0019/kWh, which was found to be lower than that of many renewable sources including that caused by biomass and photovoltaic solar panels, and was thirty times lower than coal at €0.06/kWh, or 6 cents/kWh, with the energy sources of the lowest external environmental and health costs associated with it being wind power at €0.0009/kWh.

Biofuel Use

Biofuel is defined as solid, liquid or gaseous fuel obtained from relatively recently lifeless or living biological material and is different from fossil fuels, which are derived from long-dead biological material. Various plants and plant-derived materials are used for biofuel manufacturing.

Bio-diesel

High use of bio-diesel leads to land use changes including deforestation.

Firewood

Unsustainable firewood harvesting can lead to loss of biodiversity and erosion due to loss of forest cover. An example of this is a 40-year study done by the University of Leeds of African forests, which account for a third of the world's total tropical forest which demonstrates that Africa is a significant carbon sink. A climate change expert, Lee White states that "To get an idea of the value of the sink, the removal of nearly 5 billion tonnes of carbon dioxide from the atmosphere by intact tropical forests is at issue.

According to the U.N. the African continent is losing forest twice as fast as the rest of the world. "Once upon a time, Africa boasted seven million square kilometers of forest but a third of that has been lost, most of it to charcoal."

Fossil Fuel Use

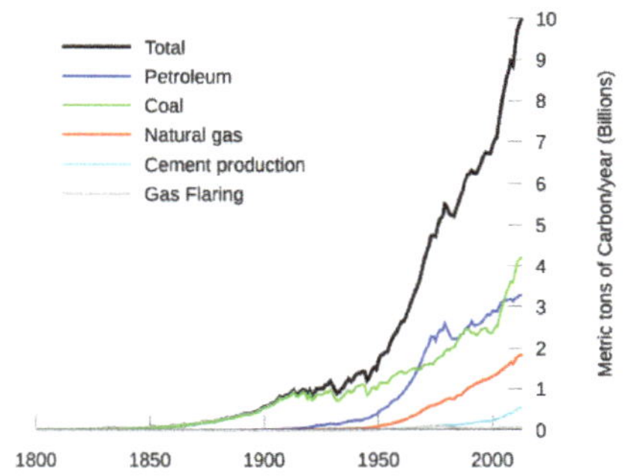

Global fossil carbon emission by fuel type, 1800-2007 AD.

The three fossil fuel types are coal, petroleum and natural gas. It was estimated by the Energy Information Administration that in 2006 primary sources of energy consisted of petroleum 36.8%, coal 26.6%, natural gas 22.9%, amounting to an 86% share for fossil fuels in primary energy production in the world.

In 2013 the burning of fossil fuels produced around 32 billion tonnes (32 gigatonnes) of carbon dioxide and additional air pollution. This caused negative externalities of $4.9 trillion due to global warming and health problems (> 150 $/ton carbon dioxide).

Carbon dioxide is one of the greenhouse gases that enhances radiative forcing and contributes to global warming, causing the average surface temperature of the Earth to rise in response, which climate scientists agree will cause major adverse effects.

Coal

The environmental impact of coal mining and burning is diverse. Legislation passed by the U.S. Congress in 1990 required the United States Environmental Protection Agency (EPA) to issue a plan to alleviate toxic pollution from coal-fired power plants. After delay and litigation, the EPA now has a court-imposed deadline of March 16, 2011, to issue its report.

Petroleum

A beach after an oil spill.

The environmental impact of petroleum is often negative because it is toxic to almost all forms of life. The possibility of climate change exists. Petroleum, commonly referred to as oil, is closely linked to virtually all aspects of present society, especially for transportation and heating for both homes and for commercial activities.

Gas

Natural gas is often described as the cleanest fossil fuel, producing less carbon dioxide per joule delivered than either coal or oil., and far fewer pollutants than other fossil fuels. However, in absolute terms it does contribute substantially to global carbon emissions, and this contribution is projected to grow. According to the IPCC Fourth Assessment Report, in 2004 natural gas produced about 5,300 Mt/yr of CO_2 emissions, while coal and oil produced 10,600 and 10,200 respectively (Figure 4.4); but by 2030, according to an updated version of the SRES B2 emissions scenario, natural gas would be the source of 11,000 Mt/yr, with coal and oil now 8,400 and 17,200 respectively. (Total global emissions for 2004 were estimated at over 27,200 Mt.)

In addition, natural gas itself is a greenhouse gas far more potent than carbon dioxide when released into the atmosphere but is released in smaller amounts.

Electricity Generation

The environmental impact of electricity generation is significant because modern society uses large amounts of electrical power. This power is normally generated at power plants that convert some other kind of energy into electrical power. Each such system has advantages and disadvantages, but many of them pose environmental concerns.

Reservoirs

The environmental impact of reservoirs is coming under ever increasing scrutiny as the world demand for water and energy increases and the number and size of reservoirs increases. Dams and the reservoirs can be used to supply drinking water, generate hydroelectric power, increasing the water supply for irrigation, provide recreational opportunities and for flood control. However, adverse environmental and sociological impacts have also been identified during and after many reservoir constructions. Whether reservoir projects are ultimately beneficial or detrimental—to both the environment and surrounding human populations— has been debated since the 1960s and probably long before that. In 1960 the construction of Llyn Celyn and the flooding of Capel Celyn provoked political uproar which continues to this day. More recently, the construction of Three Gorges Dam and other similar projects throughout Asia, Africa and Latin America have generated considerable environmental and political debate.

Nuclear Power

Nuclear power activities involving the environment; mining, enrichment, generation and geological disposal.

The environmental impact of nuclear power results from the nuclear fuel cycle, operation, and the effects of nuclear accidents.

The routine health risks and greenhouse gas emissions from nuclear fission power are smaller than those associated with coal, oil and gas. However, there is a "catastrophic risk" potential if containment fails, which in nuclear reactors can be brought about by over-heated fuels melting and releasing large quantities of fission products into the environment. The most long-lived radioactive wastes, including spent nuclear fuel, must be contained and isolated from humans and the environment for hundreds of thousands of years. The public is sensitive to these risks and there has been considerable public opposition to nuclear power.

The 1979 Three Mile Island accident and 1986 Chernobyl disaster, along with high construction costs, ended the rapid growth of global nuclear power capacity. A further disastrous release of radioactive materials followed the 2011 Japanese tsunami which damaged the Fukushima I Nuclear Power Plant, resulting in hydrogen gas explosions and partial meltdowns classified as a Level 7 event. The large-scale release of radioactivity resulted in people being evacuated from a 20 km exclusion zone set up around the power plant, similar to the 30 km radius Chernobyl Exclusion Zone still in effect.

Wind Power

Livestock grazing near a wind turbine.

The environmental impact of wind power when compared to the environmental impacts of fossil fuels, is relatively minor. According to the IPCC, in assessments of the life-cycle global warming potential of energy sources, wind turbines have a median value of between 12 and 11 (gCO_{2eq}/kWh) depending, respectively, on if offshore or onshore turbines are being assessed. Compared with other low carbon power sources, wind turbines have some of the lowest global warming potential per unit of electrical energy generated.

While a wind farm may cover a large area of land, many land uses such as agriculture are compatible with it, as only small areas of turbine foundations and infrastructure are made unavailable for use.

There are reports of bird and bat mortality at wind turbines as there are around other artificial structures. The scale of the ecological impact may or may not be significant, depending on specific circumstances. Prevention and mitigation of wildlife fatalities, and protection of peat bogs, affect the siting and operation of wind turbines.

There are anecdotal reports of negative health effects from noise on people who live very close to wind turbines. Peer-reviewed research has generally not supported these claims.

Aesthetic aspects of wind turbines and resulting changes of the visual landscape are significant. Conflicts arise especially in scenic and heritage protected landscapes.

Mitigation

Energy Conservation

Energy conservation refers to efforts made to reduce energy consumption. Energy conservation can be achieved through increased efficient energy use, in conjunction with decreased energy consumption and/or reduced consumption from conventional energy sources.

Energy conservation can result in increased financial capital, environmental quality, national security, personal security, and human comfort.Individuals and organizations that are direct consumers of energy choose to conserve energy to reduce energy costs and promote economic security. Industrial and commercial users can increase energy use efficiency to maximize profit.

Energy Policy

Energy policy is the manner in which a given entity (often governmental) has decided to address issues of energy development including energy production, distribution and consumption. The attributes of energy policy may include legislation, international treaties, incentives to investment, guidelines for energy conservation, taxation and other public policy techniques.

References

- "IPCC Working Group III – Mitigation of Climate Change, Annex II I: Technology - specific cost and performance parameters" (PDF). IPCC. 2014. p. 10. Retrieved 1 August 2014.
- "Energy Consumption: Total energy consumption per capita". Earth trends Database. World Resources Institute. Retrieved 2011-04-21.
- "Environmental impacts of coal power: air pollution". Union of Concerned Scientists. 2009. Retrieved 2011-04-22.

Permissions

All chapters in this book are published with permission under the Creative Commons Attribution Share Alike License or equivalent. Every chapter published in this book has been scrutinized by our experts. Their significance has been extensively debated. The topics covered herein carry significant information for a comprehensive understanding. They may even be implemented as practical applications or may be referred to as a beginning point for further studies.

We would like to thank the editorial team for lending their expertise to make the book truly unique. They have played a crucial role in the development of this book. Without their invaluable contributions this book wouldn't have been possible. They have made vital efforts to compile up to date information on the varied aspects of this subject to make this book a valuable addition to the collection of many professionals and students.

This book was conceptualized with the vision of imparting up-to-date and integrated information in this field. To ensure the same, a matchless editorial board was set up. Every individual on the board went through rigorous rounds of assessment to prove their worth. After which they invested a large part of their time researching and compiling the most relevant data for our readers.

The editorial board has been involved in producing this book since its inception. They have spent rigorous hours researching and exploring the diverse topics which have resulted in the successful publishing of this book. They have passed on their knowledge of decades through this book. To expedite this challenging task, the publisher supported the team at every step. A small team of assistant editors was also appointed to further simplify the editing procedure and attain best results for the readers.

Apart from the editorial board, the designing team has also invested a significant amount of their time in understanding the subject and creating the most relevant covers. They scrutinized every image to scout for the most suitable representation of the subject and create an appropriate cover for the book.

The publishing team has been an ardent support to the editorial, designing and production team. Their endless efforts to recruit the best for this project, has resulted in the accomplishment of this book. They are a veteran in the field of academics and their pool of knowledge is as vast as their experience in printing. Their expertise and guidance has proved useful at every step. Their uncompromising quality standards have made this book an exceptional effort. Their encouragement from time to time has been an inspiration for everyone.

The publisher and the editorial board hope that this book will prove to be a valuable piece of knowledge for students, practitioners and scholars across the globe.

Index